Differential Equations and Linear Algebra (0-13-086250-9) *Farlow, Hall, McDill, and West*

Differential Equations and Linear Algebra, 2e (0-13-263757-X) *Stephen Goode*

Differential Equations and Linear Algebra (0-13-011118-X) *Michael Greenberg*

Multivariable Mathematics, 4e (0-13-067276-9) *Williamson/Trotter*

Differential Equations: An Applied Approach (0-13-044930-X) *J.M. Cushing*

Differential Equations (0-13-598137-9) *Polking, Boggess, Arnold*

Differential Equations with Boundary Value Problems (0-13-091106-2) *Polking, Boggess, Arnold*

Differential Equations: Computing & Modeling, 3e (0-13-067667-4) *Edwards/Penney*

Differential Equations and Boundary Value Problems: Computing & Modeling, 3e (0-13-065245-8) *Edwards/Penney*

Elementary Differential Equations, 5e (0-13-145773-X) *Edwards/Penney*

Elementary Differential Equations with Boundary Value Problems, 5e (0-13-145774-8) *Edwards/Penney*

Differential Equations: A Systems Approach (0-13-046026-5) *Bruce Conrad*

Differential Equations and Boundary Value Problems: A Systems Approach (0-13-093419-4) *Bruce Conrad*

Differential Equations with Boundary Value Problems (0-13-015927-1) *Selwyn Hollis*

Differential Equations with Graphical and Numerical Methods (0-13-084376-8) *Bernard Banks*

Differential Equations: Modeling with MATLAB (0-13-736539-X) *Paul Davis*

Elementary Differential Equations, 8e (0-13-508011-8) *Rainville/Bedient/Bedient*

Calculus of Variations
Mechanics, Control, and Other Applications

Charles R. MacCluer
Michigan State University

Upper Saddle River, New Jersey 07458

Library of Congress Cataloging-in-Publications Data

MacCluer, Charles R.
  Calculus of variations: mechanics, control, and other applications/
    Charles R. MacCluer.
    p. cm
  Includes index.
  ISBN 0-13-142383-5
    1. Calculus of variations. I. Title.

QA315.M23 2005
515'.64—dc22   2004040069

Executive Acquisitions Editor: *George Lobell*
Editor-in-Chief: *Sally Yagan*
Vice President/Director of Production and Manufacturing: *David W. Riccardi*
Production Editor: *Debbie Ryan*
Senior Managing Editor: *Linda Mihatov Behrens*
Assistant Managing Editor: *Bayani Mendoza de Leon*
Executive Managing Editor: *Kathleen Schiaparelli*
Assistant Manufacturing Manager/Buyer: *Michael Bell*
Manufacturing Manager: *Trudy Pisciotti*
Marketing Manager: *Halee Dinsey*
Marketing Assistant: *Rachael Beckman*
Cover Designer: Bruce Kenselaar
Creative Director: *Jayne Conte*
Director of Creative Services: *Paul Belfanti*
Editorial Assistant: *Jennifer Brady*
Front Cover Image: *Susan Van Etten/Stock Boston*
Back Cover Image: *Jim Olive/Peter Arnold Inc.*

©2005 Pearson Education, Inc.
Pearson Prentice Hall
Pearson Education, Inc.
Upper Saddle River, NJ 07458

All rights reserved. No part of this book may be reproduced, in any form or by any means, without permission in writing from the publisher.

Pearson Prentice Hall[R] is a trademark of Pearson Education, Inc.

Printed in the United States of America

10 9 8 7 6 5 4 3 2 1

ISBN: 0-13-142383-5

Pearson Education, Ltd., *London*
Pearson Education Australia PTY. Limited, *Sydney*
Pearson Education Singapore, Pte., Ltd
Pearson Education North Asia Ltd, *Hong Kong*
Pearson Education Canada, Ltd., *Toronto*
Pearson Education de Mexico, S.A. de C.V.
Pearson Education - Japan, *Tokyo*
Pearson Education Malaysia, Pte. Ltd

This book is dedicated to my wife Ann.

Contents

Preface   ix
Acknowledgments   x

1 Preliminaries
1.1 Directional Derivatives and Gradients   1
1.2 Calculus Rules   3
1.3 Contour Surfaces and Sublevel Sets   5
1.4 Lagrange Multipliers   7
1.5 Convexity   9
    Exercises   12

2 Optimization
2.1 Mathematical Programming   17
2.2 Linear Programming   22
2.3 Statistical Problems   23
2.4 Variational Problems   24
    Exercises   24

3 Formulating Variational Problems
3.1 Shortest Distance between Two Points (CVP 1)   31
3.2 Graph with Least Surface of Revolution (CVP 2)   32
3.3 The Catenary (CVP 3)   32
3.4 The Brachistochrone (CVP 4)   33
3.5 Cruise-Climb (CVP 5)   34
3.6 Shapes of Minimum Resistance (CVP 6)   35
3.7 Hamilton's Principle   36
3.8 Isoperimetric Problems   39
    Exercises   40

4 The Euler-Lagrange Equation
4.1 One Degree of Freedom   45
4.2 Two Special Cases: No y, No x 48
4.3 Multiple Degrees of Freedom   52
4.4 The Hamiltonian   54
4.5 A Closer Look   58
    Exercises   60

5 Constrained Problems
5.1 Dido's Problem	67
5.2 Statement of the Problem	68
5.3 The Inverse Function Theorem	68
5.4 The Euler-Lagrange Equation for Constrained Problems	69
5.5 Example Applications	71
5.6 Multiple Degrees of Freedom	76
5.7 Nonintegral Constraints	78
5.8 Hamilton's Principle with Constraints	81
Exercises	86

6 Extremal Surfaces
6.1 A Soap Film (CVP 15)	91
6.2 Stable Flows (CVP 17)	96
6.3 Schrödinger's Equation (CVP 18)	97
6.4 Eigenvalue Problems	98
6.5 Rayleigh-Ritz Numerics	99
Exercises	104

7 Optimal Control
7.1 A Rolling Cart (OCP 1)	109
7.2 General Formulation	112
7.3 Reinvestments (OCP 2)	113
7.4 Average Voltage (OCP 3)	115
7.5 A Time-Optimal Problem (OCP 4)	116
7.6 The Bang-Bang Principle	118
7.7 The Maximum Principle	120
7.8 Example Applications	124
Exercises	129

8 The LQ Problem
8.1 Problem Statement	139
8.2 State Feedback	142
8.3 Stability	143
8.4 The LQR Problem	146
8.5 A Tracking Servo	152
Exercises	156

9 Weak Sufficiency

9.1 Weak versus Strong Extrema	161
9.2 First and Second Variations	162
9.3 In Application	166
9.4 The Integrand $p\eta'^2 + q\eta^2$	170
9.5 Weak Local Sufficiency	173
Exercises	177

10 Strong Sufficiency

10.1 The Goal	185
10.2 Flows	186
10.3 Flows of the Euler-Lagrange Equation	190
10.4 The E-Function and Strong Sufficiency	192
10.5 The Existence of Flows	196
Exercises	205

11 Corner Points

11.1 Corners and Extremals	209
11.2 First Erdmann Corner Condition	211
11.3 The Figurative	213
11.4 Second Erdmann Corner Condition	217
Exercises	221

Appendix A. The Inverse Function Theorem	223
Appendix B. Picard's Theorem	229
Appendix C. The Divergence Theorem	233
Appendix D. A MATLAB Cookbook	235
References	249
Index	251

Preface

About This Book
This book is aimed at the junior- or senior-level student of mathematics, science, and engineering. It can also be used as an amusing summer course for graduate students by a judicious use of the starred exercises and proofs. Chapters 1–7 form a leisurely undergraduate semester course.

The difficulty of the book ramps up gradually—Chapter 8 is at a strong senior level, while Chapters 9 and 10 (Weak and Strong Sufficiency) and Chapter 11 (Corner Points) are more abstract and at very strong senior or graduate level.

The charm of this subject is found in its classical applications accessible to any student with calculus. We have attempted to downplay (at first) the technical details, to instead develop technique. As a result, even a modestly equipped student can carry away a strong understanding of the subject based on practice with the calculations. The starred proofs employ advanced machinery but are sketched in an expository style that may be comprehensible to undergraduates.

Why This Book?
There is no modern text at this level that is accessible to students armed only with calculus. There are of course the fine classic Dover editions of Fox, Sagan, Weinstock, Ewing, and Gelfand/Fomin. But these books are all showing their age, and, unlike our book, none of these incorporate a simple introduction to optimal control, the bang-bang theorem, Pontryagin's maximum principle, or linear-quadratic control design. Some of the most entertaining applications of the calculus of variations are found in optimal control.

To the Instructor
At times much of the detail is thrown into the Exercises. This is to facilitate flow and better display the attractive big picture. You may include some of these solutions in your lectures or assign them in some proportion consonant with your degree of commitment to the Moore system. A disk of solutions is available upon request. Additions and corrections to the text will be updated at `http://www.math.msu.edu/~maccluer/PrenHall/additions.pdf`.

Acknowledgments

One great joy of University life is having living resources available for the mere asking. I thank my colleagues David E. Blair, William C. Brown, B-Y Chen, Leonid Freidovich, Milan Miklavcic, Boris Mordukhovich, Fedor Nazarov, Sheldon Newhouse, George Pappus, Jacob Plotkin, Clark Radcliffe, Elias Strangas, Ralph Svetic, Lal Tummala, Clifford Weil, Peter R. Wolenski, Lijian Yang, Vera Zeidan, and Zhengfang Zhou.

Many students helped shape this book. They suffered through the early write-ups and typos and often suggested valuable improvements. I thank the undergraduate students Daniel Brian Bouk, Lynne M. Evasic, Leonard Joseph Ford, Tanya Christine Hopkalo, Harold Leatherman Hunt, Rachel C. McCormick, Megan Jayne Mercer, John E. Mills, Christopher Thomas Morling, Jessica A. Munger, Jacquelyn M. Ormiston, Lindsay J. D. Radke, Stephanie L. Semann, Michael J. Stinson, Pieter C. vanRooyen, Julie K. Waibel, and Mengmeng Yu.

My special thanks go to the graduate students Michelle L. Boorom, Matthew T. Brenneman, Alberto A. Condori, Chinthaka V. Hettitantri, Ki-Moon Lee, Laura M. Stadelman, Steven W. Sy, Brian J. Vessell, and Jared Wasburn-Moses.

The reviewers of this book offered many helpful suggestions. For their insights I thank Mark Coffey of the Colorado School of Mines, Gregor Kovacic of Rensselaer Polytechnic Institute, Boris Mordukovich of Wayne State University, and Eduardo Sontag of Rutgers University.

Finally, I thank George Lobell, Executive Editor, Prentice Hall, for his continued help and support.

Charles R. MacCluer
maccluer@msu.edu

Chapter 1

Preliminaries

This chapter reviews basic tools from calculus that are used in the calculus of variations—directional derivatives, gradients, the chain rules, contour surfaces, sublevel sets, Lagrange multipliers, and the basic notion of convexity. All of these concepts form the basic toolset for attacking optimization problems.

1.1 Directional Derivatives and Gradients

A point in $\mathbb{R}^n$ is denoted by $x = (x_1, x_2, \ldots, x_n)^\top$, which is an $n \times 1$ column vector (the superscript $^\top$ denotes transpose). Suppose a function f of x represents the profit of a commercial enterprise, where x is a vector of the parameters of the operation such as labor costs, production output levels, price of the commodity, and so on. The manager naturally desires to know in which direction from the present operating point x^0 should the company move in order to obtain the maximum increase in profit. The desired direction is found by using a multi-variable notion of a derivative: For each unit vector $u \in \mathbb{R}^n$, the *directional derivative* of f at x^0 in direction u is given by

$$D_u f(x^0) = \lim_{h \to 0^+} \frac{f(x^0 + hu) - f(x^0)}{h}, \tag{1.1}$$

provided the limit exists. Geometrically, this limit is the slope of the line tangent to the curve above x^0 obtained by cutting the hypersurface $z = f(x)$ with the hyperplane determined by u and the z-axis. See Figure 1.1. The directional derivatives in the directions parallel to the coordinate axes are, of course, the familiar *partial derivatives*

$$D_u f(x^0) = \frac{\partial f(x^0)}{\partial x_k},$$

when $u = (0, 0, \ldots, 0, 1 \ (k\text{th position}), 0, \ldots, 0)^\top$.

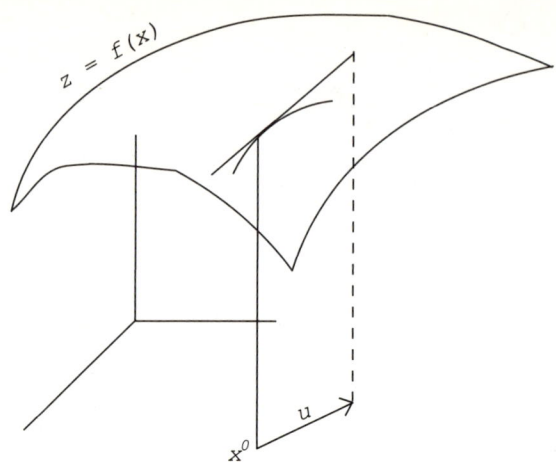

Figure 1.1 The graph of y = f(x) is cut by a plane perpendicular to the coordinate space in the direction u. The slope of the line tangent to the curve thus cut out is given by the directional derivative $D_u f(x^0)$.

The function f is *differentiable* at x^0 provided it is linearly approximated by its tangent plane near x^0. This means there exists a constant (row) vector $a = (a_1, a_2, \ldots, a_n)$ such that for all x in some open ball $B = \{x;\ |x - x^0| < r\}$ of radius r about x^0,

$$f(x) = f(x^0) + a(x - x^0) + \epsilon(x) \qquad (1.2a)$$
$$= f(x_1^0, x_2^0, \ldots, x_n^0) + a_1(x_1 - x_1^0) + a_2(x_2 - x_2^0) + \cdots + a_n(x_n - x_n^0) + \epsilon(x),$$

where $\epsilon(x)$ is such that

$$\lim_{x \to x^0} \frac{\epsilon(x)}{|x - x^0|} = 0. \qquad (1.2b)$$

If f is differentiable at x^0, the row vector a is called the *gradient* of f at x^0, is necessarily unique (Exercise 1.1), and is denoted by $\nabla f(x^0)$.

Directional derivatives may exist in all directions without a function being differentiable (Exercise 1.2). However, there is a simple formula for the directional derivative in terms of the gradient when f is differentiable.

Theorem A. Suppose f is differentiable at x^0. The directional derivative in the direction u is obtained as

$$D_u f(x^0) = \nabla f(x^0) u, \qquad (1.3)$$

1.2 Calculus Rules

where the gradient is calculated by

$$\nabla f(x^0) = \left(\frac{\partial f}{\partial x_1}, \frac{\partial f}{\partial x_2}, \ldots, \frac{\partial f}{\partial x_n}\right)\bigg|_{x^0}. \tag{1.4}$$

Proof. Exercise 1.3.

The kth partial derivative $\partial f/\partial x_k$ represents the sensitivity of f to changes in the kth component variable x_k.

Corollary A. For $u \in \mathbb{R}^n$ with $|u| = 1$,

$$D_u f(x^0) = |\nabla f(x^0)| \cos \theta, \tag{1.5}$$

where θ is the angle between (the unit vector) u and $\nabla f(x^0)$.

Corollary B. The gradient $\nabla f(x^0)$ points in the direction of maximum increase of f at x^0. This maximal rate of increase is $|\nabla f(x^0)|$.

Example 1. Let $f(x, y) = x^2 + xy + y^3$. The directional derivative of f at $(1, 2)$ in the direction $u = (a, b)$ is then

$$D_{(a,b)} f(1, 2) = (2x + y, x + 3y^2)|_{(1,2)} \begin{vmatrix} a \\ b \end{vmatrix} = 4a + 13b,$$

with maximal directional derivative (in the gradient direction) of value $\sqrt{16 + 169}$.

1.2 Calculus Rules

The product formula (1.3) for the directional derivative is actually an instance of a much more general result, the chain rule.

Theorem B. (The First Chain Rule) Suppose that each component of the vector curve $x = x(t)$ is differentiable at $t = t^0$, and that $f = f(x)$ is differentiable at $x^0 = x(t^0)$. Then

$$\frac{d\,f(x(t))}{dt}\bigg|_{t=t^0} = \nabla f(x^0) \dot{x}(t^0) = \sum_{k=1}^{n} \frac{\partial f}{\partial x_k}\bigg|_{x=x^0} \frac{dx_k}{dt}\bigg|_{t=t^0}. \tag{1.6}$$

In more transparent notation,

$$\frac{d}{dt} f(x(t)) = \sum_{k=1}^{n} \frac{\partial f}{\partial x_k} \frac{dx_k}{dt}.$$

Proof. Combine $x = x^0 + (t - t^0)\dot{x}(t^0) + \epsilon$ with equation (1.2) (Exercise 1.5).

Example 2. Let $f(x, y) = x^2 - 2xy + y^3$ and $x(t) = \cos t$, $y(t) = \sin t$. Then

$$\frac{d}{dt} f(x(t), y(t)) = -(2\cos t - 2\sin t)\sin t + (-2\cos t + 3\sin^2 t)\cos t.$$

The First Chain Rule itself is a special case of the following more general rule.

Theorem C. (The Second Chain Rule) Suppose $x = x(u) \in \mathbb{R}^n$ is differentiable at $u = u^0$ where $u = (u_1, \ldots, u_r)^\top$ (that is, each component of x is differentiable), and suppose also that $f = f(x)$ is differentiable at $x^0 = x(u^0)$. Then at $u = u^0$,

$$\frac{\partial f}{\partial u_k} = \sum_{j=1}^{n} \frac{\partial f}{\partial x_j} \frac{\partial x_j}{\partial u_k}. \qquad (1.7)$$

Proof. Exercise 1.5.

Example 3. Let $f(x, y) = x^2 - 2xy + y^3$, $x = u^2 - v^2$, $y = u^2 + v^2$. Then

$$\frac{\partial f}{\partial v} = (2x - 2y)(-2v) + (-2x + 3y^2)(2v)$$

$$= -4v^2(-2v) + (-2u^2 + 2v^2 + 3(u^2 + v^2)^2)(2v).$$

Corollary. Suppose $f(x) = (f_1(x), f_2(x), \ldots, f_m(x))$, where $x = x(u)$. Then, where differentiable, we have the matrix relation

$$\left[\frac{\partial f_i}{\partial u_j}\right] = \left[\frac{\partial f_i}{\partial x_j}\right] \left[\frac{\partial x_i}{\partial u_j}\right].$$

That is,

$$\frac{\partial f_i}{\partial u_j} = \sum_{k=1}^{n} \frac{\partial f_i}{\partial x_k} \frac{\partial x_k}{\partial u_j}. \qquad (1.8)$$

1.3 Contour Surfaces

Theorem D. (The Mean Value Theorem) Suppose x^0 and x^1 belong to $\mathbb{R}^n$ and $f : \mathbb{R}^n \to \mathbb{R}$ is differentiable at each point on the line segment $[x^0, x^1] = \{tx^1 + (1-t)x^0;\ 0 \le t \le 1\}$. Then there exists a point $x^t \in [x^0, x^1]$ so that

$$f(x^1) - f(x^0) = \nabla f(x^t)(x^1 - x^0).$$

Proof. Apply the usual one-dimensional mean value theorem and the chain rule to $t \mapsto f(tx^1 + (1-t)x^0)$ defined on $[0, 1]$.

1.3 Contour Surfaces and Sublevel Sets

The locus of all points x satisfying $f(x) = f(x^0)$ is called the *contour surface* (or a *contour curve* if $n = 2$) of f through the point $x = x^0$.

Under mild assumptions on the differentiability of f near $x = x^0$, we may generically, in theory, solve for one of the components x_j of x, say for x_n, in terms of the remaining x_k [i.e., $x_n = x_n(x_1, \ldots, x_{n-1})$ so that the portion of the contour surface $f(x) = f(x^0)$ near $x = x^0$ is the graph of $z = f(x_1, x_2, \ldots, x_{n-1}, x_n(x_1, \ldots, x_{n-1}))$ near $(x_1^0, \ldots, x_{n-1}^0)$].

Example 4. Consider the contour curve $x^2 - y^2 = 1$ of $f(x, y) = x^2 - y^2$ through the point $(1, 0)$.

The locus has two disjoint branches—one in the first and fourth quadrant, the other in the second and and third quadrant. But only one branch passes through $(1, 0)$, where we may solve for x in terms of y:

$$x = \sqrt{1 + y^2},$$

valid for all y. See Figure 1.2.

The technical result that validates the preceding intuition in the general case is a workhorse of mathematics, the so-called *Implicit Function Theorem*. This theorem is easily deducible from another workhorse, the so-called *inverse function theorem*. See Chapter 4 and Appendix A. See also Exercises 1.15 and 5.13.

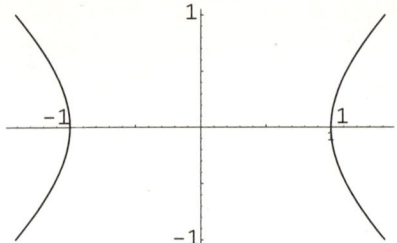

Figure 1.2 The contour curve of $x^2 - y^2 = 1$.

The value $f(x(t))$ along any curve $C : x = x(t)$ passing through $x^0 = x(0)$ that lies within the contour surface $f(x) = f(x^0)$ must, of course, be constantly $f(x^0)$. But then by the first chain rule,

$$\frac{d}{dt}f(x(t)) = \frac{d}{dt}f(x^0) = 0 = \nabla f(x(t))\dot{x}(t), \tag{1.9}$$

where of course $v = \dot{x}$ is tangent to the curve C given by $x = x(t)$ (i.e., the gradient is normal to the curve C). But since C was an arbitrary curve in the contour surface $f(x) = f(x^0)$,

The gradient is normal to the contour surface.

(See Exercise 1.38.) This means precisely that the hyperplane $\mathcal{H}$ that is tangent to the contour surface

$$S = \{x : f(x) = f(x^0)\}$$

at $x = x^0$ has normal vector $\nabla f(x^0)$, and hence the points of $\mathcal{H}$ are those x that satisfy the equation

$$\nabla f(x^0)(x - x^0) = 0. \tag{1.10}$$

Example 6. The plane tangent to the unit sphere $f(x, y, z) = x^2 + y^2 + z^2 = 1$ at $(2\sqrt{3}/5, 2/5, 3/5)$ has equation

$$(x - 2\sqrt{3}/5, \ y - 2/5, \ z - 3/5) \cdot (4\sqrt{3}/5, 4/5, 6/5) = 0,$$

that is,

$$\sqrt{3}x + 10y + 15z = 25.$$

1.4 Lagrange Multipliers

Example 7. The plane tangent to the surface $z = f(x, y)$ at (x_0, y_0) is the plane tangent to the contour surface $z - f(x, y) = 0$ at (x_0, y_0, z_0), which must be perpendicular to the gradient $(-f_x, -f_y, 1)$ and so has equation
$$z - z_0 = (x - x_0)f_x(x_0, y_0) + (y - y_0)f_y(x_0, y_0).$$
Compare with (1.10).

A *sublevel set* of f is of the form $\{x : f(x) \leq c\}$ and can be thought of as the side of a contour surface that contains the points where f is smaller than the values on the surface. The preceding considerations (specifically Corollary B) show that the gradient direction at x^0 points outward from the sublevel set formed by setting $c = f(x^0)$.

1.4 Lagrange Multipliers

The following result is a recurring motif of the calculus of variations.

Theorem E. Suppose both f and g are realvalued functions of the real vector variable $x = (x_1, \ldots, x_n)^\top$, $n > 1$, with continuous first partial derivatives on an open neighborhood of $x = x^0$. Suppose also that

(a) $x = x^0$ is a local minimum or maximum of f on the contour surface $g(x) = g(x^0)$ and

(b) $\nabla g(x^0) \neq 0$.

Then, for some scalar λ,
$$\nabla f(x^0) + \lambda \nabla g(x^0) = 0. \tag{1.11}$$

Geometric Intuition. Unless $\nabla f(x^0)$ is normal to the contour surface $g(x) = g(x^0)$, by (1.5) there are curves through x^0 lying within the contour surface along which f has both larger and smaller values than $f(x^0)$. See Figure 1.3.

Formal Proof. If $\nabla f(x^0) = 0$, then (1.11) holds trivially with $\lambda = 0$. So assume the result false, that ∇f and ∇g are not multiples of one another at $x = x^0$. But then we may renumber the variables until the row 2-vectors $(\partial f/\partial x_1, \partial f/\partial x_2)$ and $(\partial g/\partial x_1, \partial g/\partial x_2)$ are not multiples at $x = x^0$ (Exercise 1.17). Consider the mapping
$$(x_1, x_2, \ldots, x_n)^\top \longmapsto (f(x), g(x), x_3, x_4, \ldots, x_n)^\top, \tag{1.12}$$

continuously differentiable in a neighborhood of $x = x^0$. The Jacobian

$$\det \begin{bmatrix} f_{x_1}(x^0) & f_{x_2}(x^0) & \cdot & \cdot & f_{x_n}(x^0) \\ g_{x_1}(x^0) & g_{x_2}(x^0) & \cdot & & g_{x_n}(x^0) \\ 0 & 0 & 1 & 0 & \cdot & 0 \\ 0 & 0 & 0 & 1 & 0 & \cdot \\ \cdot & \cdot & & \cdot & & \\ \cdot & \cdot & & \cdot & & \\ 0 & 0 & \cdot & \cdot & 0 & 1 \end{bmatrix} = \det \begin{bmatrix} f_{x_1}(x^0) & f_{x_2}(x^0) \\ g_{x_1}(x^0) & g_{x_2}(x^0) \end{bmatrix} \quad (1.13)$$

of this mapping is nonzero, and so by the inverse function theorem of §5.3 and Appendix A, there are neighborhoods of x^0 and its image on which the mapping (1.12) is bijective with a continuous inverse. But this means we can choose points x near x^0 where $g(x) = g(x^0)$ but where $f(x)$ is larger and smaller than $f(x^0)$.

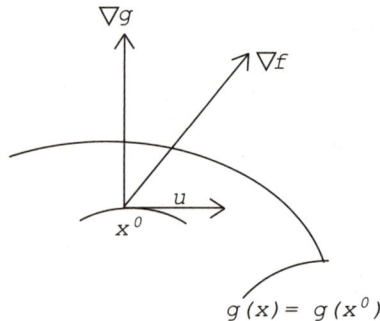

Figure 1.3 Unless the gradients are parallel, by (1.5) there will be nearby points on the contour where f is both larger and smaller than its value at x^0.

Example 8. What point on the unit sphere $x^2 + y^2 + z^2 = 1$ is closest to the point $(2, 1, 1)$?

Of course we can predict that the answer lies along the line from the given point to the origin, namely the point $p = (2, 1, 1)/\sqrt{6}$. Let us obtain this via Lagrange multipliers.

We are to minimize $f(x, y, z) = (x-2)^2 + (y-1)^2 + (z-1)^2$ subject to the constraint $g(x, y, z) = x^2 + y^2 + z^2 = 1$. By the theorem,

$$\frac{\partial}{\partial x}(f + \lambda g) = 2(x - 2) + 2\lambda x = 0, \quad (1.14\text{a})$$

1.5 Convexity

$$\frac{\partial}{\partial y}(f + \lambda g) = 2(y - 1) + 2\lambda y = 0, \quad (1.14b)$$

$$\frac{\partial}{\partial z}(f + \lambda g) = 2(z - 1) + 2\lambda z = 0. \quad (1.14c)$$

Thus

$$x = \frac{2}{1+\lambda}, \quad y = \frac{1}{1+\lambda}, \quad z = \frac{1}{1+\lambda}. \quad (1.15)$$

But then by the constraint,

$$x^2 + y^2 + z^2 = \frac{4}{(1+\lambda)^2} + \frac{1}{(1+\lambda)^2} + \frac{1}{(1+\lambda)^2} = 1,$$

and so, as predicted, $1 + \lambda = \sqrt{6}$.

1.5 Convexity

For $x^0, x^1 \in \mathbb{R}^n$ and $0 \le t \le 1$, let

$$x^t = tx^1 + (1-t)x^0 = x^0 + t(x^1 - x^0). \quad (1.16)$$

A set $C \subseteq \mathbb{R}^n$ is *convex* provided that the line segment connecting any two points in C is itself in C (i.e., whenever $x^0, x^1 \in C$, then $x^t \in C$ for all $t \in [0,1]$). A function $f : \mathbb{R}^n \to \mathbb{R}$ is *convex* provided that

$$f(x^t) \le tf(x^1) + (1-t)f(x^0) \quad (1.17)$$

for all $x^0, x^1,$ and $t \in [0, 1]$.

An important aspect of convexity is that the tangent plane not only approximates the function *locally* but in fact sits below the function *globally*.

Theorem F. Suppose f is convex and is differentiable at x^0. Then for all $x \in \mathbb{R}^n$,

$$f(x) \ge f(x^0) + \nabla f(x^0)(x - x^0). \quad (1.18)$$

Proof. Fix $x = x^1 \in \mathbb{R}^n$ and let x^t be as in (1.16). Then by (1.17),

$$f(x^t) = f(x^0 + t(x^1 - x^0)) \le tf(x^1) + (1-t)f(x^0),$$

which can be rearranged to

$$\frac{f(x^0 + t(x^1 - x^0)) - f(x^0)}{t} \leq f(x^1) - f(x^0)$$

Letting $t \searrow 0$ implies (1.18).

Corollary A. If $\nabla f(x^0) = 0$ and f is convex, then f has a global minimum at $x = x^0$.

Thus convexity plays a special role in minimization—*the first-order necessary condition that the gradient vanishes is sufficient under convexity.* This will be a recurring theme of this book. For example, Theorem E has a convexity converse.

Corollary B. Suppose f and g are both convex and differentiable at x^0. Consider the problem of minimizing f on the contour $g(x) = g(x^0)$. If there exists $\lambda \geq 0$ so that

$$\nabla f(x^0) + \lambda \nabla g(x^0) = 0,$$

then x^0 is a solution to the problem—that is, $f(x) \geq f(x^0)$ for all x satisfying $g(x) = g(x^0)$.

Geometrical interpretation. The minimum value in the optimization problem is the infimum over c that satisfy

$$\{x : f(x) \leq c\} \bigcap \{x : g(x) = 0\} \neq \emptyset.$$

Since sublevel sets of convex functions are convex (Exercise 1.29), if the gradients of f and g at x^0 are pointing in opposite directions, these level sets must be tangential, and so the sets in (1.19) will become disjoint if $c < f(x^0)$.

Proof of Corollary B. Since $\lambda \geq 0$, the function $h(x) = f(x) + \lambda g(x)$ is convex on $\mathbb{R}^n$ (Exercise 1.28). By Corollary A, $f(x) + \lambda g(x) \geq f(x^0) + \lambda g(x^0)$ for all $x \in \mathbb{R}^n$. In particular, then, $f(x) \geq f(x^0)$ for all x satisfying $g(x) = g(x^0)$.

The final result of this chapter is a set analogue to the function result in Theorem F and is fundamental in optimization. It in effect says that closed convex sets can be separated by a hyperplane from any points not in it.

1.5 Convexity

Theorem G. (The Separation Theorem) Suppose C is closed and convex and $x^0 \notin C$. Then there exists $\zeta \in \mathbb{R}^n$ and $r \in \mathbb{R}$ so that

$$\langle \zeta, c \rangle < r < \langle \zeta, x^0 \rangle \qquad \text{for all } c \in C. \tag{1.19}$$

Thus a closed convex set C can be separated from a point x^0 not in it by a hyperplane $\{x; \langle \zeta, x \rangle = r\}$.

Proof.[1] Suppose $x^0 \notin C$, and let $c^0 \in C$ be such that

$$|x^0 - c^0| = \inf\{|x^0 - c|;\ c \in C\}.$$

Such a c^0 exists since C is closed. Set $\zeta = x^0 - c^0$, and fix r with $\langle \zeta, c^0 \rangle < r < \langle \zeta, x^0 \rangle$, the latter existing since $\langle \zeta, x^0 - c^0 \rangle = |\zeta|^2 > 0$. The theorem will be easily derived from the following claim:

$$\langle \zeta, c^1 - c^0 \rangle \leq 0 \qquad \text{for all } c^1 \in C. \tag{1.20}$$

To prove (1.20), let $c^1 \in C$, c^t be as in (1.16), and θ the angle between ζ and $c^1 - c^0$. Of course, θ is also the angle between ζ and $c^t - c^0$. Since c^0 is the closest point in C to x^0, we have by the law of cosines

$$|\zeta|^2 = |x^0 - c^0|^2 \leq |x^0 - c^t|^2 = |\zeta|^2 + |c^t - c^0|^2 - 2|\zeta| \cdot |c^t - c^0| \cos\theta$$

$$= |\zeta|^2 + t^2|c^1 - c^0|^2 - 2t|\zeta| \cdot |c^1 - c^0| \cos\theta.$$

Rearranging terms and dividing by $0 < t \leq 1$ gives

$$2|\zeta| \cos\theta \leq t|c^1 - c^0|.$$

Letting $t \searrow 0$ implies $\cos\theta \leq 0$ and consequently that the claim (1.20) is valid.

To finish the proof, we simply note from (1.20) that for each $c \in C$,

$$\langle \zeta, c \rangle \leq \langle \zeta, c^0 \rangle < r < \langle \zeta, x^0 \rangle,$$

which is (1.19).

---

[1] The *inner product* $\langle \zeta, x \rangle = \zeta^\top x$ is in this case the ordinary dot product.

Exercises

1.1 Prove that the row vector a in (1.2a) is unique.

1.2 Show that $f(x) = |x|$ is not differentiable at $x^0 = 0$ yet its directional derivative exists in every direction from $x^0 = 0$.

1.3 Prove the formula (1.3) for computing the directional derivative by combining (1.1) with (1.2).

1.4 Compute the directional derivative of $f = x^3 + 4xy^2 + 3y^2 + z^5$ at $(1, -1, 1)$ in the direction $(2, 2\sqrt{3}, 3)/5$.

1.5 Prove Theorems B and C.

1.6 For polar coordinates $x = r\cos\theta$, $y = r\sin\theta$, compute the Jacobians
$$a = \det A = \det \begin{pmatrix} x_r & x_\theta \\ y_r & y_\theta \end{pmatrix}$$
and
$$b = \det B = \det \begin{pmatrix} r_x & r_y \\ \theta_x & \theta_y \end{pmatrix}.$$
What is their matrix product AB?

1.7 Compute $\rho_\phi = \partial\sqrt{x^2 + y^2 + z^2}/\partial\phi$ in three ways for the spherical coordinates $x = \rho\cos\theta\sin\phi$, $y = \rho\sin\theta\sin\phi$, $z = \rho\cos\phi$, as follows:

(a) Make the substitutions and then take the partial derivative.

(b) Use the second chain rule.

(c) Without computation via a geometric interpretation.

1.8 Find the plane tangent to the surface $z = x^2 - 2xy + y^3$ above $(x, y) = (1, -1)$.

Answer: $z = 4x + y - 1$.

1.9 Parametrize the contour curve $y^2 - 2xy - x^2 = -2$ near $(1, 1)$ by solving for one variable in terms of the other. Why is it possible to solve for one in terms of the other but not *vice versa*?

Hint: Look up the statement of the implicit function theorem in Appendix A. Or, graph the equation.

Exercises

1.10 Show that $u(x,t) = f(x-ct) + g(x+ct)$ satisfies the *wave equation*
$$\frac{\partial^2 u}{\partial t^2} = c^2 \frac{\partial^2 u}{\partial x^2}$$
for any sufficiently smooth f and g.

1.11 Conversely, suppose u satisfies the wave equation of Exercise 1.10. Set $v(\alpha, \beta) = u(x,t)$, where $\alpha = x - ct$ and $\beta = x + ct$. Show that
$$\frac{\partial^2 v}{\partial \alpha\, \partial \beta} = 0.$$
Deduce that u is of the form $u(x,t) = f(x-ct) + g(x+ct)$.

1.12 Find the line tangent to the contour curve at $(1,0)$ of Example 4 in two ways—first using the parametrization $x = \sqrt{1+y^2}$, second via equation (1.10).

1.13 Find the tangent plane to the surface $z = x^3 - 2xy^2 + 5y^4$ at $(3,-1)$.

1.14 Find the hyperplane tangent to the contour surface $f(w,x,y,z) = x^2 + yz + 2xw^2 + z^2 = 5$ at $(1,-1,1,2)$. Locally parametrize this surface in terms of x, y, z.

1.15 State the implicit function theorem (Exercise 4.13) for the special case $f(x,y) = c$ in several of the preceding examples.

1.16 The *Laplacian* is the operator $\nabla^2 u = u_{xx} + u_{yy} + u_{zz}$. Show that $\nabla^2 r = 2/r$ and $\nabla^2(1/r) = 0$ when $r = \sqrt{x^2 + y^2 + z^2}$. What are the corresponding results when $r = \sqrt{x^2 + y^2}$?

1.17 Verify that resubscripting the variables can indeed render the Jacobian (1.13) nonzero (Recall that row rank equals column rank).

1.18 Find the extreme values of $f(x,y,z) = x^2 + 2y^2 + z^2$ subject to $g(x,y,z) = 2x + 2y + z = 13$. Verify that these values are indeed extreme.

1.19 What point on the surface $2x^2 + y^4 + z^2 = 1$ is farthest from the origin?

1.20 Using Lagrange multipliers, prove that a $n \times n$ symmetric matrix A has at least one real eigenvalue.

Hint: Maximize $f(x) = \langle Ax, x \rangle = x^T A x$ subject to $g(x) = \langle x, x \rangle = 1$.

1.21 Prove that if all partials of f are continuous on an open set Ω, then f is differentiable at every point of Ω.

1.22 Prove that if f is differentiable at a point p, it is continuous at p.

1.23 Give an example or prove no example exists of a function $z = f(x, y)$ that is

(a) continuous at $(0,0)$ but not differentiable there

(b) continuous and both partials exist at $(0,0)$ yet is not differentiable there

(c) nowhere differentiable

(d) nowhere continuous but where both partials exist at infinitely many points (x, y)

(e) differentiable at $(0,0)$ but nowhere else.

1.24 If $z = f(x, y)$ is everywhere continuous in x and y separately, must f be continuous?

1.25 Show that the composition of differentiable functions is again differentiable.

1.26 Prove the **theorem on Lagrange multipliers for multiple constraints:** Suppose f, g_i, $i = 1, 2, \ldots, r < n$ are real-valued functions of the real vector variable $x = (x_1, \ldots, x_n)^T$, with continuous first partial derivatives on an open neighborhood of $x = x^0$. Suppose also that

(a) $x = x^0$ is a local minimum or maximum of f on the intersection of the r contour surfaces $g(x) = g_i(x^0)$, and that

(b) The matrix $[\nabla g_i(x^0)]$ is nonsingular.

Then for some r scalars λ_i,

$$\nabla f(x^0) + \sum_{i=1}^{r} \lambda_i \nabla g_i(x^0) = 0.$$

Hint: Find an independent set among the r rows ∇g_i. Apply the proof of Theorem D.

1.27 Show that any local minimum of a convex function f on a convex set C is in fact a global minimum on all of C.

1.28 Suppose f and g are convex and $\mu \geq 0$. Show that $f + g$ and μf are also convex.

1.29 Show that every sublevel set of a convex function is a convex set, but that the converse is false.

1.30 The *epigraph* epi f of a function $f : \mathbb{R}^n \to \mathbb{R}$ is defined as epi $f = \{(x, r) \in \mathbb{R}^{n+1} : r \geq f(x)\}$. Show that f is a convex function if and only if epi f is a convex set.

1.31 Suppose $g : \mathbb{R} \to \mathbb{R}$ is convex and $x < y < z$. Show that

$$\frac{g(z) - g(y)}{z - y} \geq \frac{g(y) - g(x)}{y - x}.$$

Deduce that for any (not necessarily differentiable) convex function $f : \mathbb{R}^n \to \mathbb{R}$, the directional derivative $D_u f(x)$ exists for all x and u.

How is this compatible with the fact that $f(x) = |x|$ is convex?

1.32 (Converse of Theorem F) Suppose f is differentiable on $\mathbb{R}^n$, and (1.18) holds for each $x \in \mathbb{R}^n$. Show that f is convex.

1.33 Suppose $f : \mathbb{R}^n \to \mathbb{R}$ is everywhere differentiable. Show that f is convex if and only if

$$(\nabla f(x) - \nabla f(y))(x - y) \geq 0$$

for all $x, y \in \mathbb{R}^n$. If f is C^2 (i.e., if all second partial derivatives of f are everywhere continuous), show that f is convex if and only if its *Hessian*

$$D^2 f(x) = \left[\frac{\partial^2 f}{\partial x_i \partial x_j}\right]$$

is positive semidefinite. (A square matrix A is *positive semidefinite* if the quadratic form $Q(x) = x^\top A x \geq 0$ for all x.)

1.34 Show that a closed set is convex if and only if it is the intersection of all closed halfspaces that contain it.

1.35 Suppose $C \subseteq \mathbb{R}^n$ is closed and convex, and c^0 belongs to the boundary of C. Show that there exists a vector $\zeta \in \mathbb{R}^n$ so that
$$\langle \zeta, c - c^0 \rangle \leq 0 \quad \text{for all } c \in C.$$
Hint: Consider points x^i not in C satisfying $x^i \to c^0$, and let $\zeta^i = (x^i - c^i)/|x^i - c^i|$, where $c^i \in C$ is the closest point to x^i.

1.36 Let $E \subseteq \mathbb{R}^n$. The *convex hull* conv E of E is the set of all convex combinations of points in E, that is,
$$\operatorname{conv} E = \left\{ \sum_{i=1}^{k} t_i e_i : k > 0,\ e_i \in E,\ \sum_{i=1}^{k} t_i = 1, t_i \geq 0 \right\}.$$
Show that conv E is the minimal convex set containing E.

1.37* (Caratheodory's Theorem) Show that if $E \subseteq \mathbb{R}^n$ is closed, then in the definition of conv E given in Exercise 1.36, the convex combinations need be at most of length $k = n + 1$. Go by induction on n.

1.38 Prove more carefully than in §1.3 via the implicit function theorem that *the gradient at a nonsingular point is perpendicular to the contour surface through that point.*

1.39 Is the point c^0 of nearest approach to x^0 in the proof of the separation theorem (Theorem G) unique?

1.40* Assume that each particle of a system can exist at only one of the discrete energy levels $E_1 < E_2 < E_3 < \cdots$. Let p_n be the frequency with which a particle is found at level E_n, giving that $1 = p_1 + p_2 + p_3 + \cdots$. Fix the mean energy of the system $E = E_1 p_1 + E_2 p_2 + E_3 p_3 + \cdots$. Show that when *entropy* $H = -p_1 \log p_1 - p_2 \log p_2 - p_3 \log p_3 - \cdots$ is maximum, there is a unique *temperature* T so that each $p_n = e^{-E_n/T}/Q$, where the *partition function* $Q = e^{-E_1/T} + e^{-E_2/T} + e^{-E_3/T} + \cdots$, and hence $H = E/T + \log Q$ ([Feynman]).

Hint: Use Lagrange multipliers.

Chapter 2
Optimization

Many if not most problems of commerce, government, and warfare are *optimization problems* of the following type:

Find the maximum (or minimum) value of the *cost function* $J = J(x)$ subject to a collection C of constraints on the variables x.

For example, how do we maximize profit given labor and material costs? What container holds the maximum volume for a given surface area? What is the correct angle to mount a sonar transducer to optimize submarine detection? What is the best air search pattern for finding shipwreck survivors? What price will maximize revenue? What cam shape maximizes available power? What cluster bomb spacing induces maximal damage to enemy airfields? What gives the most bang for the buck?

Many of these problems are nowadays classed as problems of *operations research* (OR), a field begun by mathematicians, almost immediately abandoned to other disciplines once the field was determined to be useful and profitable. In this chapter we work through typical concrete optimization problems solvable by mathematical programming, linear programming, and the Monte Carlo method. We will see why certain optimization problems of an important type require a distinct approach—the calculus of variations.

2.1 Mathematical Programming

Mathematical programming problems are optimization problems of a familiar type:

Find the maximum (or minimum) of the cost function $J = J(x)$ subject to the constraints $g(x) \leq 0$ and $h(x) = 0$.

Approach. If the cost function $J = J(x)$ is a smooth function of a finite number of real variables $x = (x_1, x_2, \ldots, x_n)$, where the variables x are only constrained to lie in some region Ω, then we have seen in calculus that interior local extreme values of J will occur at *critical (stationary) points* of J, namely at interior points where the n partials $\partial J/\partial x_i$ all simultaneously vanish (i.e., where the gradient $\nabla J = 0$). Often this means that one is left merely with checking the value of J at some finite list of critical points (candidate extrema) and on the boundary $\partial \Omega$ of Ω.

Problem A. What is the maximum value taken on by

$$f(x) = 2x^3 - 9x^2 + 12x - 2 \qquad (2.1)$$

on the interval $[-4, 3]$?

Solution 1. Interior extrema occur at critical points, where $f'(x) = 6x^2 - 18x + 12 = 6(x-1)(x-2) = 0$, giving the two candidates $x = 1, 2$ for interior extrema. The values of f at these critical points and the two end points $x = -4, 3$ are displayed in Table 2.1.

Table 2.1 The values of the f of (2.1) at critical points and endpoints.

x	-4	1	2	3
$f(x)$	-322	3	2	7

Because *continuous functions on closed and bounded (compact) subsets of $\mathbb{R}^n$ are bounded, and in fact achieve their maximum and minimum values,* we know that the maximum value of f is obtained somewhere on $[-4, 3]$, either at the endpoints or at an interior critical point where $f'(x) = 0$ or does not exist. Thus we need only search among the four values of f in Table 2.1. Hence the maximum value of f is 7, achieved at the right endpoint $x = 3$. Also see the crucial Exercise 2.16.

Solution 2. Graph the function. Pick off the extrema. For example, via MATLAB (Appendix D):

```
% graphing routine
x = -4:0.1:3;
plot(x,2*x.*x.*x -9*x.*x + 12*x -2)
```

Or with *Mathematica*,

```
f[x_] := 2 x x x - 9 x x + 12 x -2;
Plot[f[x],{x,-4,3}]
```

2.1 Mathematical Programming

Solution 3. Instruct a computer to search through 'all' values. Pseudocode for such a routine would look like

```
a = -4, b = 3, dx = 0.01
xmax = a, fmax = -1000
x = a
loop
  if f(x) > fmax, then xmax = x and fmax = f(x)
  x = x + dx
until x > b
write xmax, fmax
```

Critique. Unfortunately, in practical problems, the function we wish to maximize or minimize is most often of the form $f = f(x, \alpha)$, where α is a vector of parameters whose values are uncertain and known only within certain bounds; thus the last two solution methods fail. The first approach will also fail when the region is no longer closed or bounded—for example when $f(x) = e^{-x}$ on $(0, \infty)$. More detailed analysis is warranted. Computer search can obtain the results only to the resolution specified and is easily humbled by cost functions of many variables.

Problem B. What is the maximum value of

$$J = e^{-x^2 - y^2 + 2y} \tag{2.2}$$

and where does it occur?

Solution.

$$\frac{\partial J}{\partial x} = -2xe^{-x^2-y^2+2y} \quad \text{and} \quad \frac{\partial J}{\partial y} = -(2y-2)e^{-x^2-y^2+2y}$$

are both simultaneously 0 exactly when $2x = 0$ and $2y - 2 = 0$. Thus the only candidate for a local extremum is the point $(0, 1)$ where $J = e$. But is this the minimum or a saddle point rather than the maximum? We could go on to the second derivative test but it is far easier to complete the square (Exercise 2.3). We have indeed found the maximum value of J.

Problem C. How should 80 rods (1320 feet) of fencing be used to enclose a rectangular meadow that borders a river?

Solution 1. Introduce the variables x, y as shown in Figure 2.1. The objective (cost) function to be maximized is the area

$$J(x, y) = xy, \qquad (2.3\text{a})$$

subject to the perimeter constraints

$$C(x, y) = x + 2y = 80, \quad x, y \geq 0. \qquad (2.3\text{b})$$

Eliminate the variable y by means of the constraint $y = 40 - x/2$ to obtain the new single-variable cost

$$J = 40x - x^2/2$$

to be maximized over the interval $0 \leq x \leq 80$. Proceeding as in the first solution to Problem A, we see that the only critical point is $x = 40$ and that the objective vanishes at the endpoints. Thus the meadow should have dimensions $x = 40, y = 20$ in order to maximize the grazing area.

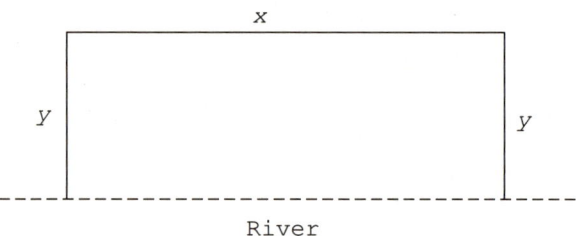

Figure 2.1 A meadow is fenced on three sides.

Solution 2. The function $J = xy$ maximizes only where its differential dJ vanishes:

$$y \, dx + x \, dy = 0. \qquad (2.4\text{a})$$

But the differential of the (constant) constraint is also 0:

$$dx + 2 \, dy = 0. \qquad (2.4\text{b})$$

For the system (2.4) to have a solution, its determinant must vanish:

$$\det \begin{bmatrix} y & x \\ 1 & 2 \end{bmatrix} = x - 2y = 0,$$

2.1 Mathematical Programming

that is, the aspect ratio $x{:}y$ must be 2 to 1. Thus $x = 40$ since $x + 2y = 80$.

Put more geometrically, suppose the surface $z = J(x,y)$ peaks above (say) the point (x_0, y_0), and so has a horizontal tangent plane at this peak with equation $dz = y_0 \, dx + x_0 \, dy$ in the local coordinate system mounted at the peak: $dx = x - x_0$, $dy = y - y_0$, $dz = z - J(x_0, y_0)$. Because this tangent plane is horizontal at extrema, $dz \equiv 0$. But we are constrained to look for points along the line $d80 = 0 = dx + 2\,dy$. This geometric argument is formalized in the next famous solution method.

Solution 3. (Via Lagrange Multipliers) Recall two facts about directional derivatives:

The gradient of f at p points in the direction (from p) of maximal increase of f.

The gradient of f at p is normal to the contour surface $f = f(p)$ through p.

So unless the objective's contour line $J(x,y) = J(x_0, y_0)$ through (x_0, y_0) is parallel at (x_0, y_0) to the constraint's contour line $C(x,y) = C(x_0, y_0)$ through (x_0, y_0), there must be neighboring points on the constraint contour where J is larger (and smaller) than the value $J(x_0, y_0)$. That is, at a local extrema, ∇J and ∇C must be scalar multiples of one another. See Figure 2.2. So the *Method of Lagrange Multipliers* (Theorem D, §1.4) is the execution of the following three steps:

Step 1. Form the function $F = J - \lambda C$.

Step 2. Take and set equal to 0 each of the partials of F to obtain the system

$$\nabla J = \lambda \nabla C \quad \text{and} \quad C = 0. \tag{2.5}$$

Step 3. Solve the system (2.5).

For Problem C, the system (2.5) becomes

$$(y, x) = \lambda(1, 2) \quad \text{and} \quad x - 2y = 80. \tag{2.6}$$

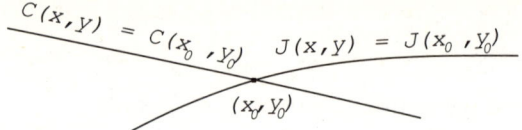

Figure 2.2 Unless the tangents are parallel at the point of intersection, the cost J will increase at nearby points along the constraint contour.

Solution 4. Introduce a "virtual" meadow as shown in Figure 2.3. The area of the resulting rectangle for a given perimeter of 160 must maximize when the rectangle becomes a square (i.e., when $x = 40$).

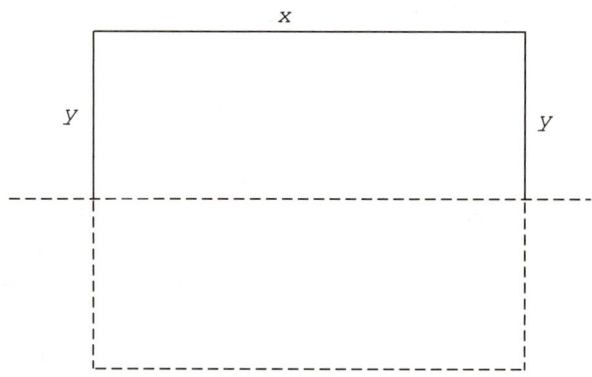

Figure 2.3 A virtual fence is added to form a rectangle.

2.2 Linear Programming

In this class of problems, the cost and the constraint functions are linear.

Problem D. Find the minimum value of $J(x, y) = x - 2y$ provided that x, y satisfy the constraints $x + y \geq 1$, $y \leq 2$, $x \leq y + 1$, and $x \leq 2$.

Solution. The *feasible* points (i.e., the points satisfying the constraints), form a polygon—in this problem, the 4-gon shown in Figure 2.4.

2.3 Statistical Problems

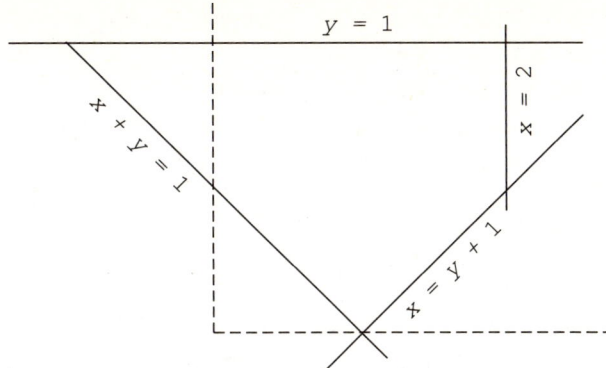

Figure 2.4 The feasible points of Problem D.

Because the cost function J is linear, it is impossible that all of its partials can vanish simultaneously somewhere. Thus there are no interior extrema.

Extrema must occur on the boundary.

But J is a linear function along each line segment making up the boundary of the set of feasible points and so takes on its extreme points at endpoints.

The extrema occur at the vertices.

So to find the maximum or minimum of the cost J, we 'only' need check the values at the vertices of the polygon of feasible points (Exercise 2.6). Unfortunately, in the important problems of industry or warfare, there may be (say) 200 variables with 40 constraints, yielding a possible $200!/40!160! \approx 2 \times 10^{42}$ vertices to check—a more efficient algorithm is essential. The search for improved linear programming algorithms is a central problem of commerce. See [Pierre] and [MacCluer, 2000].

2.3 Statistical Problems

Many optimization problems are basically statistical in nature, such as optimizing the response of servers to requests for service, elevator algorithms, anticipating and lengthening times to failure and financial risk reduction. One of the most famous is an artillery problem disguised as an innocent business problem [Morse and Kimball].

Problem E. (The newsboy problem) A newsboy sells newspapers outside Grand Central Station. He has on average 100 customers per day. He buys papers for 50 cents each, sells them for 75 cents each, but cannot return unsold papers for a refund. How many papers should he buy? (Answer: 96 papers—see [MacCluer, 2000].)

Such problems may have analytic solutions, but more often it is enough to instruct a computer to perform many simulated experiments and use the resulting statistical data to resolve the problem. This is the powerful *Monte Carlo* method illustrated in the last several exercises below. Before trying these exercises, glance through Appendix D, a MATLAB cookbook.

2.4 Variational Problems

Problem F. What is the shortest distance between two points? That is, what path $x = x(t)$, $y = y(t)$ will minimize the distance

$$J = \int ds = \int v \, dt = \int_0^1 \sqrt{\dot{x}(t)^2 + \dot{y}(t)^2} \, dt \qquad (2.7)$$

between two given points $p_0 = (x(0), y(0))$ and $p_1 = (x(1), y(1))$?

This is an optimization problem of an entirely different character. It is clear that the cost is the arclength J between the two points p_0 and p_1. It is also clear that the constraint is that the path must begin at p_0 and end at p_1. But what are the variables? They are the various possible paths! Our variables x, y are themselves functions of a parameter t. This sort of problem requires the *calculus of variations*, the central topic of this book.

EXERCISES

2.1 Find two positive numbers x and y whose sum is minimum but whose product is 2. What are the cost, variables, and constraints of this problem?

2.2 A 10-cubic-foot oven with a square base loses heat per unit area from its top at twice the rate it loses heat from its sides, and at 4 times the rate it loses heat from its bottom. Find the dimensions of the most energy-efficient such oven.

Exercises

2.3 Complete the square in Problem B to find where the function (2.2) maximizes.

2.4 Solve Problem B by graphing with MATLAB. Use `mesh`.

2.5 Formulate an interesting optimization problem. Carefully describe your objective (cost) function J, your variables x, and the constraints.

2.6 Solve Problem D by checking the values of the cost at the vertices of Figure 2.4.

2.7 Solve Problem D without computation.

Approach: Because ∇J points in the direction of maximal increase in J, merely locate the vertex farthest upstream.

2.8 Run the MATLAB routine of the second solution of Problem A.

2.9 Where does $f(x) = x^{5/7} - 3.013x^3 + 2.7x^{1/2} - 5.09x + 1.978$ take on its maximum and minimum value on the interval $[0, 2]$?

2.10 Perform a computer search to find the maximum value of $f(x, y) = x \cos y - y \cos x$ on $0 \leq x, y \leq \pi$.

2.11 Find the minimum and maximum values taken on by $J(x, y, z) = \sqrt{(x-2)^2 + y^2 + z^2}$ on the unit sphere $x^2 + y^2 + z^2 = 1$. First solve by interpreting the problem geometrically. Then solve via Lagrange multipliers.

Answer: min $J = 1$, max $J = 3$.

2.12 Find the point on the unit sphere $x^2 + y^2 + z^2 = 1$ closest to the plane $x + 2y + 3z = 4$.

Hint: The line segment of minimum length connecting the two surfaces is normal to both.

2.13 The stiffness of a rectangular beam is proportional to the product of its width with the cube of its depth. Find the stiffest beam that can be hewn from a cylindrical log of radius R.

2.14 (Project) We must lift a mass m vertically by means of a cam rotating about point O that pushes against a lever articulated at P (see the following figure). What camshaft shape $r = r(\theta)$ yields the smoothest action? Formulate "smoothest action." For example, does this mean that the rate of change of torque about O with respect to θ should be more or less constant during the lift?

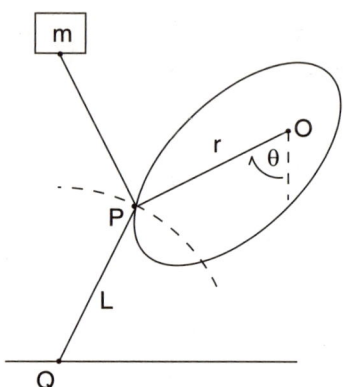

2.15 (Project) Prove that if x^0 yields a minimum value $J(x^0)$ of the linear programming problem $J(x) = c^T x$ subject to $Ax \geq b$ and $x \geq 0$, then there exists a maximal value $J^*(y^0)$ to the *dual* problem $J^*(y) = b^T y$ subject to $A^T y \leq c$ and $y \geq 0$.

Outline: Prove and apply a version of Exercise 1.26 known as the *Kuhn-Tucker* theorem.

2.16 Suppose the real-valued function f of the real-valued variable x can be expanded in a Taylor series

$$f(x) = f(a) + \frac{f'(a)}{1!}(x-a) + \frac{f''(a)}{2!}(x-a)^2 + \frac{f'''(a)}{3!}(x-a)^3 + \cdots$$

within some open neighborhood of $x = a$. Suppose f is non-constant on this neighborhood. State and prove a **necessary and sufficient condition that** $x = a$ **be a local minimum of** f in terms of the derivatives of f at $x = a$.

Answer: $f'(a) = f''(a) = \cdots = f^{(2k-1)}(a) = 0$ and $f^{(2k)}(a) > 0$ for some integer $k \geq 1$.

The remaining exercises practice the Monte Carlo method. We recommend employing MATLAB and its functions rand and randn. See Appendix D.

2.17 Write a script that computes the number $d(n)$ of positive integral divisors of an integer n. Estimate the average of $d(n)$ of the integers n between 1 and N experimentally via Monte Carlo by randomly selecting some n. Check your experimental result against actual fact.

2.18 Using Monte Carlo, estimate the integral

$$I = \int_0^1 x^2 \, dx.$$

Experiment by varying the number of trials N. Does accuracy continue to improve? Graph accuracy against the number of trials.

Outline: Repeatedly throw a dart at the rectangle $0 \leq x, y \leq 1$. The ratio of successes (where the the resulting impact (x, y) satisfies $y \leq x^2$) to the number of trials is approximately the area under the curve.

2.19 Using Monte Carlo, estimate the integral

$$I = \int_0^\pi \int_0^\pi \sin^6 xy \, dx \, dy.$$

Compare to the value obtained via `dblquad`.

2.20 Simulate a drunkard's walk. As he leaves the bar at $(0,0)$, each step is in a random direction of length $L = 1$. Estimate the expected (average) value of the square $r(N)^2$ of his distance $r(N)$ from the bar after N steps. Estimate the expected value of $r(N)$.

Exact answer: $E[r(N)^2] = N$.

Outline: After the kth step,

$$(x_k, y_k) = (x_{k-1}, y_{k-1}) + (\cos\theta, \sin\theta),$$

where θ is chosen randomly from the interval $[0, 2\pi]$.

2.21 Let X be the random variable

$$X = \frac{X_1 + X_2 + \cdots + X_N}{\sqrt{N}},$$

where X_k is the outcome from the kth flip of a fair coin where heads yields the value 1 and tails -1. Show that the expected values $E[X] = \mu = 0$ and $\sigma^2 = E[X^2] = 1$. But how is X distributed? Show experimentally using `hist` that for large N, X appears to be normally distributed, that is,

$$prob(a < X < b) = \frac{1}{\sqrt{2\pi}\sigma} \int_a^b e^{(x-\mu)^2/2\sigma^2} \, dx.$$

You will be substantiating the celebrated *central limit theorem* stated in any book on statistics.

2.22 In the same vein as in the previous exercise, an approximately normal random variable with $\mu = 0$, $\sigma = 1$ is often constructed by adding together 12 independent, uniformly distributed random variables with support $[0, 1]$ and subtracting 6. Construct such a random variable using MATLAB's uniformly distributed `rand` and check the resulting density function for normality.

2.23 A product is composed of three components and will fail if both component 1 and 2 fail or if component 3 fails. The component lifetimes are normally distributed with means $\mu = E[X] = 1, 2, 3$ and standard deviations $\sigma = 0.2, 0.5, 0.7$ respectively. Find the mean time before failure of the construct. What is the standard deviation $\sigma = E[(X - \mu)^2]^{1/2}$ of this failure time?

2.24 An integer between 1 and N is chosen randomly. What is the probability that it is the sum of two squares? Obtain this estimate experimentally.

2.25 Consider all functions f from a set X of cardinality m to a set Y of cardinality $n \leq m$. Experimentally estimate what percentage of the f are surjective (onto) for various m and n.

2.26 Experimentally verify that with probability 1, all square real matrices are invertible.

Exercises

2.27 Let X be normally distibuted with mean $\mu = 0$ and standard deviation $\sigma = 1$. Using MATLAB's `randn`, display an experimentally obtained probability distribution function of the random variable $Y = X^2$.

2.28 Experimentally verify the *prime number theorem:* The number $\pi(x)$ of primes $p \leq x$ is asymptotically $x/\log x$. Check your experimental result using MATLAB's `primes`.

2.29 Every day you must give your dog one half of a heartworm pill. You begin the mosquito season with a large bottle of N whole pills. Each day you decant an object from the bottle: If it is a half pill, give it to the dog; if whole, break it in half, administer one half to the dog, and return the other half to the bottle. After k days, what is the expected number of unbroken pills remaining?

2.30 (**Project**) Experimentally verify *Lévy's theorem*: Suppose $\zeta(x, y)$ is specified to be either 0 or 1 at each point on the boundary of the bounded domain Ω in the plane. Let $u(x, y)$ denote the probability that a particle released at (x, y) in Ω, thereafter undergoing Browian motion, will exit through the boundary at a point with ζ-value 1. Then $u(x, y)$ is a harmonic function that agrees with $\zeta(x, y)$ on the boundary of Ω.

2.31 (**Project**) State-space systems that evolve in time are often modeled using their *proper orthogonal decomposition* (POD) (or *autocorrelation*) modes ϕ_k:

$$X(t) = c_1(t)\phi_1 + c_2(t)\phi_2 + \cdots + c_m(t)\phi_m.$$

The POD modes ϕ_k are determined statistically as follows: At the jth sample time t_j, measure the state

$$X_j = (x_1(t_j), x_2(t_j), \ldots, x_m(t_j))^T$$

of the system, an $m \times 1$ column of readings. Take n samples with $n \gg m$. Then form the $m \times m$ matrix

$$A = \frac{1}{n}\Phi\Phi^T \approx E[X \cdot X^T],$$

where $\Phi = (x_i(t_j))$ is the $m \times n$ matrix of state columns X_j. The m eigenvectors ϕ_i belonging to the (nonnegative) eigenvalues of A (the squares of the singular values of Φ) form an orthogonal basis for state space.

Argue that these POD modes form the natural basis for the time evolving states of the system since they arise from statistical data of actual performance. Now take the other side and argue that the POD modes are irrelevant since they are not based on physical law—after all, you could adjoin (say) a commodity price as an additional $(m+1)$th state.

Experiment with a phase space proper orthogonal decomposition of a simulated one-mass, two-spring, or a two-mass, three-spring mechanical system. Add noise with MATLAB's `randn`, and compute the autocorrelation modes ϕ_i. Do you recover physically meaningful eigenvalues and modes?

2.32 (Project) Simulate Young's *double slit experiment* via Monte Carlo. Demonstrate that as the number of released photons grows, the impacts on the screen (behind the double slit) build up a histogram similar to an interference pattern, as if each photon were a wave interfering with itself.

2.33 (Benford's Law) In any table of physical constants, more entries lead with the digit 1 than any other digit.

Test this experimentally by constructing a table of pseudodata with entries $y = f(\texttt{rand})$, where (say) $f(x) = (3x + x^3)^2$.

Can you explain this phenomenon?

2.34 Solve the Newsboy Problem E of §2.3 via Monte Carlo simulation.

Chapter 3
Formulating Variational Problems

In this chapter we practice formulating variational problems by setting up several of the classic problems—shortest arclength, least surface of revolution, hanging cable, brachistochrone, cruise-climb, and shapes of minimum resistance. The fundamental Hamilton's principle is formulated for the spring-mass system and two pendulum systems. Finally, we formulate several isoperimetric problems. We will solve each of these problems in Chapters 4 and 5.

3.1 Shortest Distance between Two Points (CVP 1)

CVP 1. What is the shortest path between two given points in the Euclidean plane?

Formulation. Without loss of generality we may place one point at the origin $(0,0)$, the other at $(a,0)$. Consider any continuously differentiable function $y = f(x)$, $0 \leq x \leq a$, with $f(0) = 0 = f(a)$. See Figure 3.1. Our goal is to discover which such f has a graph of minimal arclength

$$J = \int_0^a ds = \int_0^a \sqrt{1 + y'^2}\, dx. \tag{3.1}$$

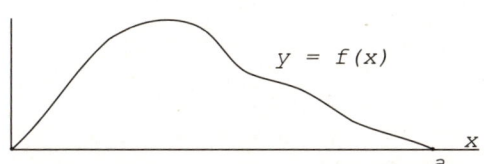

Figure 3.1 A path $y = f(x)$ from $(0, 0)$ to $(a, 0)$.

3.2 Graph with Least Surface of Revolution (CVP 2)

CVP 2. Revolve the portion of the curve $y = f(x)$ from (a, c) to (b, d) about the y-axis. Which function f yields the least surface area? See Figure 3.2.

Formulation. Our objective is to minimize the surface area

$$J = \int_a^b 2\pi x \, ds = 2\pi \int_a^b x\sqrt{1 + y'^2} \, dx. \tag{3.2}$$

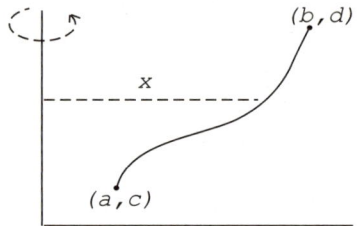

Figure 3.2 A curve $y = f(x)$ is revolved about the y-axis to produce a surface of least area.

3.3 The Catenary (CVP 3)

CVP 3. Suspend a string of length γ between two points. What shape will the string take?

Formulation. Each infinitesimal length ds of the string of Figure 3.3 contributes $dW = \rho g y \, ds$ of potential energy (referenced against $y = 0$) to the string's total potential energy of

$$J = \rho g \int_{-a}^{a} y \, ds = \rho g \int_{-a}^{a} y\sqrt{1 + y'^2} \, dx, \tag{3.3}$$

where $y = y(x)$ is the shape of the hanging string, ρ the density of the string per unit length, g the acceleration due to gravity, and ds the increment of arclength. The string at rest will have found the shape of least potential energy J.

3.4 The Brachistochrone (CVP 4)

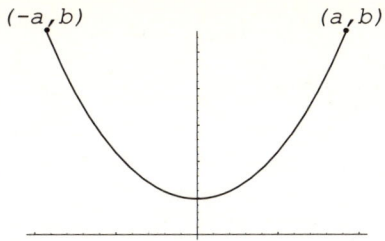

Figure 3.3 The *catenary*, the shape of least potential energy.

3.4 The Brachistochrone (CVP 4)

This is the problem that launched the calculus of variations—its name is a transliteration of βράχιστος (shortest) χρόνος (time). This problem was posed by Johann Bernoulli and G. W. von Leibniz in 1695 to smoke out their competition [Katz].

CVP 4. What is the shape of a (frictionless) slide from one place to a lower one that yields the fastest transit time?

Formulation. The kinetic energy $mv^2/2$ of the sliding body of mass m at each moment must equal the potential energy lost from the initial height. That is, using the coordinate system of Figure 3.4,

$$\frac{mv^2}{2} = mgy,$$

and so

$$v = \sqrt{2gy}. \tag{3.4}$$

The total time T needed to slide down to the terminal position (b, d) is

$$T = \int \frac{\text{distance}}{\text{rate}} = \int \frac{ds}{v} = \int_0^b \frac{\sqrt{1+y'^2}}{\sqrt{2gy}}\, dx. \tag{3.5}$$

Our objective is to find the curve $y = y(x)$ of least time T.

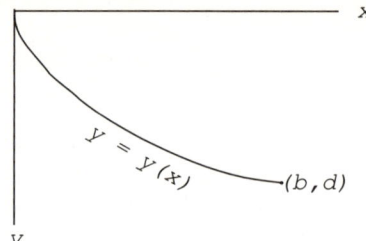

Figure 3.4 A slide is built from (0, 0) down to (b, d). Which slide shape will yield the least travel time?

3.5 Cruise-Climb (CVP 5)

CVP 5. Fighters are launched from an aircraft carrier. What climb strategy will allow the longest cruise time at patrol altitude?

Formulation. The rate of fuel consumption at cruise is constant while the total amount of fuel initially on board is known. Thus to maximize cruise time on patrol we must minimize the fuel expended while climbing to patrol altitude. As a first cut, assume a constant speed v throughout. We assume constant engine efficiency and we ignore the effect of the lightening load as fuel is burned off. At each point of the climbing path $x = x(y)$, the fuel consumed dF during time dt depends of the attack angle θ. Thus it is reasonable that dF is of the form $dF = f(\sin\theta)\,dy$. Thus our task is to minimize

$$J = \int_0^a dF = \int_0^a f(\sin\theta)dy$$

$$= \int_0^a f\left(\frac{dy}{\sqrt{dx^2 + dy^2}}\right)dy = \int_0^a f\left(\frac{1}{\sqrt{(dx/dy)^2 + 1}}\right)dy. \qquad (3.6)$$

There is a vast literature on this problem as befits its extreme importance (Exercise 3.26).

3.6 Shapes of Minimum Resistance (CVP 6)

CVP 6. (I. Newton) What surfaces of revolution have least drag?

Formulation. Suppose the curve $y = f(x)$ with $0 \leq x \leq a$ is revolved about the y-axis. The resulting solid is then dragged through a fluid at a moderate velocity v in the direction of the y-axis. See Figure 3.5. At first approximation, the resistive force exerted by each fluid particle depends on the square of velocity, since the work done overcoming this force is converted into the particle's kinetic energy. The mass of particles placed into motion by a revolved portion of arclength ds depends on the frontal area $dA = 2\pi x \cos\theta\, ds$, with those particles receiving a velocity normal to the surface of $v\cos\theta$, where θ is the angle between the direction of v and the normal to the curve (again see Figure 3.5). But $\cos\theta = dx/ds$. Thus to minimize total drag we must minimize the integral

$$J = \int 2\pi x (v\cos\theta)^2 \cos\theta\, ds$$

$$= 2\pi v^2 \int_0^a x \left(\frac{dx}{ds}\right)^3 ds = 2\pi v^2 \int_0^a \frac{x\, dx}{1 + y'^2}. \tag{3.7}$$

We are tacitly assuming that $x = x(s)$ is a monotonically increasing function—that no portion of the curve shelters other portions from the current.

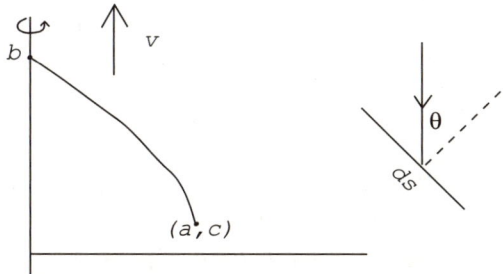

Figure 3.5 A curve is revolved about the *y*-axis and the resulting surface is pulled through a fluid at a velocity *v* parallel to the axis of revolution. The fluid resistance on a portion *ds* of the surface is reduced by $\cos^3\theta$.

3.7 Hamilton's Principle

The following is one of the most important insights of science. We will think of it as the fundamental axiom of mechanical systems. The principle predicts the nonrelativistic motion of mechanical systems from galaxies to billiard balls.

Hamilton's Principle. A conservative mechanical system, as it evolves during the time from t_1 to t_2, will choose a trajectory $q = q(t)$ that yields a *stationary* value of

$$J = \int_{t_1}^{t_2} L \, dt = \int_{t_1}^{t_2} (T - V) dt, \tag{3.8}$$

where the *Lagrangian* $L = T - V$ is the system's kinetic energy T less its potential energy V at time t.

This special curve $q = q(t)$ yields a *stationary* value of J when infinitesimal perturbations of $q = q(t)$ yield at worst only second-order changes in the value of J. That is, in the expansion

$$J(\epsilon) = \int_{t_1}^{t_2} L(t, q(t) + \epsilon \eta(t), \dot{q}(t) + \epsilon \dot{\eta}(t)) \, dt$$

$$= J(0) + J'(0)\epsilon + J''(0)\epsilon^2/2! + \cdots, \tag{3.9}$$

we must have $J'(0) = 0$ whenever $\eta(t_1) = 0 = \eta(t_2)$. See Figure 3.6. Thus the trajectory $q = q(t)$ taken by the mechanical system is a *critical* curve for the cost J. These critical curves need not minimize the cost J (Exercise 3.20–3.21). Much more on this later.

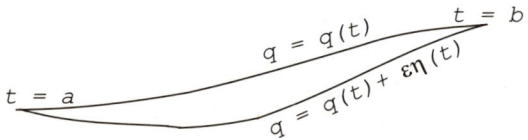

Figure 3.6 The path $q = q(t)$ is perturbed to a nearby path $q = q(t) + \varepsilon\eta(t)$ to yield a new cost $J(\varepsilon)$.

3.7 Hamilton's Principle

CVP 7. Vibrations of a spring-mass system.

Formulation. A mass m is free to slide along a horizontal frictionless rail (see Figure 3.7) restrained by an ideal spring of spring constant k. Recall that Hooke's law states that the restoring force F of the spring is proportional to the displacement q from its natural length (i.e., $F = -kq$). Hence if the mass is displaced q from rest, the work done must be

$$V = -\int F\, dq = \int_0^q kx\, dx = kq^2/2. \tag{3.10}$$

Since kinetic energy $T = m\dot{q}^2/2$, the spring-mass system must move by Hamilton's principle in such a way as to yield a stationary value of

$$J = \int_{t_1}^{t_2} L\, dt = \frac{1}{2}\int_{t_1}^{t_2} (m\dot{q}^2 - kq^2)\, dt. \tag{3.11}$$

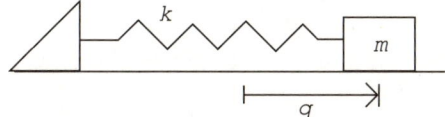

Figure 3.7 A spring-mass system.

CVP 8. The planar pendulum.

Formulation. A bob of mass m is connected by a massless rigid rod of length a to a point of pivot which constrains the resulting pendulum's motion to one plane. See Figure 3.8.

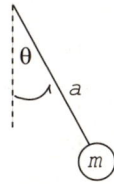

Figure 3.8 Planar pendulum.

The kinetic energy of the bob is $T = mv^2/2 = I\omega^2/2 = a^2m\dot{\theta}^2/2$, while its potential energy is $V = $ weight $\times$ height $= mga(1 - \cos\theta)$.

Thus the trajectories of motion make stationary the integral

$$J = \int_{t_1}^{t_2} L\, dt = \int_{t_1}^{t_2} \left(\frac{a^2 m \dot\theta^2}{2} - mga(1 - \cos\theta)\right) dt. \qquad (3.12)$$

CVP 9. The articulated planar pendulum.

Formulation. A bob of mass m is connected by two rigid articulated, massless rods to a pivot (see Figure 3.9).

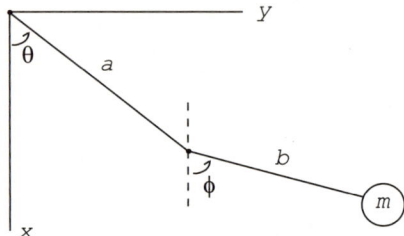

Figure 3.9 The articulated planar pendulum.

The coordinate of the mass at the tip is the sum of two vectors:

$$(x, y) = a(\cos\theta, \sin\theta) + b(\cos\phi, \sin\phi), \qquad (3.13)$$

giving a tip velocity of

$$(\dot x, \dot y) = a\dot\theta(-\sin\theta, \cos\theta) + b\dot\phi(-\sin\phi, \cos\phi). \qquad (3.14)$$

Thus the kinetic energy of this system is

$$T = \frac{m}{2}[(a\dot\theta \sin\theta + b\dot\phi \sin\phi)^2 + (a\dot\theta \cos\theta + b\dot\phi \cos\phi)^2]. \qquad (3.15)$$

The potential energy is the work needed to lift m above bottom dead center $(a+b, 0)$, namely,

$$V = mg(a + b - x) = mg(a + b - a\cos\theta - b\cos\phi). \qquad (3.16)$$

Therefore, the equations of motion are obtained from finding stationary values of the integral

$$J = \int_{t_1}^{t_2} (T - V)\, dt. \qquad (3.17)$$

3.8 Isoperimetric Problems

Note that this model requires the two *degrees of freedom* $q_1 = \theta$ and $q_2 = \phi$.

Summary. Hamilton's principle reduces the modeling of conservative mechanical systems to a standard procedure:

Step 1. Write kinetic energy T and potential energy V in terms of the configuration variables q and $\dot{q}$.
Step 2. Write down the cost integral J of the Lagrangian $L = T - V$.
Step 3. Impose the Lagrange equations (as we will explain in Chapter 4).
Step 4. Solve the resulting ordinary differential equations to obtain the motions of the system.

3.8 Isoperimetric Problems

CVP 10. (Queen Dido of Carthage)[1] Granted a portion of a coastline of Africa as border, what is the largest country that can be enclosed by a given remaining perimeter?

Formulation. Say the coastline is the line segment $[a, b]$ and the interior of the country is bounded by the graph of $y = y(x) \geq 0$ with $y(a) = 0 = y(b)$. The goal is to maximize the area

$$J = \int_a^b y \, dx \qquad (3.18a)$$

subject to the fixed perimeter constraint

$$\int_a^b \sqrt{1 + y'^2} \, dx = \gamma. \qquad (3.18b)$$

---

[1] According to myth, after her husband was killed by her brother Pygmalion, Dido fled Tyre with her retinue (and a sizable portion of the Tyre treasury) to Mediterranean Africa. There she purchased from a naive king all the land *that could be enclosed by the hide of an ox*. After slicing the hide into thin strips and tying end to end, she enclosed a sizeable parcel that became the city-state of Carthage, *circa* 853 BC. Virgil in his *Aeneid* translates the myth 300 years back in time to the Fall of Troy in order to interweave her story with that of Aeneas.

CVP 3. The Catenary (reprise) A careful reexamination of the hanging string of §3.3 reveals it to be an isoperimetric problem.

Formulation. We must minimize potential energy

$$J = \rho g \int_{-a}^{a} y\sqrt{1 + y'^2}\, dx \qquad (3.19\text{a})$$

subject to the constant arclength requirement

$$\int_{-a}^{a} \sqrt{1 + y'^2}\, dx = \gamma. \qquad (3.19\text{b})$$

EXERCISES

3.1 Reformulate CVP 1 in polar coordinates.

3.2 Reformulate CVP 2 when the curve is rotated about the x-axis.

3.3 Formulate the problem of finding the shortest path between any two points in space.

3.4 We wish to move a mass m against heavy steady wind from $x = 0$ to $x = 100$ in $T = 10$ seconds. Formulate the problem of finding the speed $v = v(t)$ that requires the least work to be done in this task.

Hint: Air resistance is proportional to the square of relative velocity. Work is the time integral of the product of force with velocity.

3.5 A slab of glass is bounded by the planes $x = 0$ and $x = 1$. The speed of light entering from the left and exiting from the right is affected by a nonconstant index of refraction of the material so that its velocity c is a function of x. Formulate the problem of finding the path taken by a light ray through the glass.

Fermat's Principle: Light chooses the path of least time.

3.6 State and formulate an interesting variational problem.

Exercises

3.7 Find the Lagrangian of the following multiple mass-spring system:

![mass-spring system diagram with masses $m_1, m_2, \ldots, m_n$, springs $k_0, k_1, k_2, \ldots, k_{n-1}, k_n$, and coordinates $q_1, q_2, \ldots, q_n$]

3.8 Suppose a particle at (x, y) moving in the plane experiences an acceleration $a = kr$ toward the origin, where $r = \sqrt{x^2 + y^2}$. Using Hamilton's principle, formulate what orbits are possible for this particle.

3.9 Suppose a particle at (x, y) moving in the plane experiences an acceleration $a = k/r^2$ toward the origin, where $r = \sqrt{x^2 + y^2}$. Using Hamilton's principle, formulate what orbits are possible for this particle.

3.10 Write down the Lagrangian for two point masses M and m in space, say for the Earth and the Moon. You will need 6 degrees of freedom.

3.11 (Project) The Earth-Moon system is not conservative—because of the Moon-induced tides on Earth, the system is losing mechanical energy. (Think of the sound of surf.) Find the system's total energy $E = T + V$ neglecting tides by including the rotational kinetic energy of both objects. Now introduce tidal losses. Is the Moon approaching the Earth or receding?

3.12 Find the Lagrangian for the double pendulum:

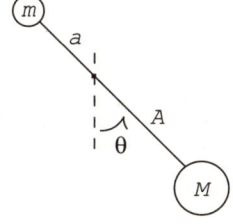

3.13 (Project) Formulate the problem of finding the optimal path of a surface-to-air missile as it rises to intercept an unsuspecting Airbus™. Take into account air resistance, the decreasing mass of the missile as it burns off fuel, and so on.

3.14 (Project) Carefully reconstruct and report on the history of the brachistochrone problem. Read original papers and comment on their contents.

3.15* Suppose the linear operator A on a real inner product space X is *self-adjoint* (i.e., $\langle Au, v \rangle = \langle u, Av \rangle$ for all u, v in X). Suppose also that the operator A is *positive semidefinite* (i.e., $\langle Au, u \rangle \geq 0$). Show that for any fixed f in X,

$$J(u) = \langle Au, u \rangle - 2\langle f, u \rangle$$

is a convex function of u in X. Thus any local minimum of J is global.

3.16* Suppose in the Exercise 3.15 that the inner product space is $X = \mathbb{R}^n$. Show that J has a (unique) minimum at some $u = u_0$ and that $Au_0 = f$, provided that A is *positive definite* (i.e., $\langle Au, u \rangle > 0$ for all $u \neq 0$).

Outline: J is bounded below because $\langle Au, u \rangle \geq \lambda \langle u, u \rangle$, where λ is the least eigenvalue of A. But J takes on its minimum value only where $\nabla J = 0$.

3.17* Find a bounded self-adjoint, positive definite operator A on ℓ^2 where

$$J(u) = \langle Au, u \rangle - 2\langle f, u \rangle$$

is not bounded below.

Hint: Take $A = \text{diag}(1/k^2)$, $2f = \{1/k^{3/4}\}$, $u^{(n)} = ne_n$.

3.18* Show that for the differential operator $Au = -u''$,

$$J(u) = \langle Au, u \rangle - 2\langle f, u \rangle = -\int_0^1 u''(x)u(x)\,dx - \int_0^1 2f(x)u(x)\,dx$$

$$= -\int_0^1 u'(x)^2 + 2f(x)u(x)\,dx,$$

for $u(0) = u(1) = 0$, is bounded below.

Exercises

3.19 (**Project**) Compare the relative contributions of Euler and Lagrange. Read their original papers concerning the calculus of variations and comment on their contents.

3.20 Hamilton's principle is sometimes incorrectly stated as follows: *A mechanical system will choose a trajectory that minimizes J, the integral of the Lagrangian L.*

Show this statement is not correct even for the spring-mass system CVP 7.

Outline: Consider the simplest case $m = k = 1$, $t_1 = 0$, $t_2 = T \neq n\pi$. Show J has no minimum or maximum value, by using the actual mechanical trajectories $x = a\sin(t - \phi)$.

3.21 Hamilton's principle is at other times stated as follows: *Once the initial and terminal configurations $q(t_1)$, $q(t_2)$ are specified, a mechanical system will select the trajectory that extremizes (3.8), the integral J of the Lagrangian.*

Show that this statement of Hamilton's principle is not correct if the word *extremizes* means "maximizes or minimizes."

Outline: Use the previous Lagrangian $L = \dot{q}^2 - q^2$, $q(0) = 0 = q(T)$ with $\pi < T < \sqrt{10}$. Show that the cost J at $q = at(T - t)$ can be made arbitrarily large while, on the other hand, the cost can be made arbitrarily large in the negative direction by taking $q = a\sin(t\pi/T)$. The cost J has no extreme values.

3.22 Hamilton's principle is sometimes incorrectly stated as follows: *A mechanical system will choose a trajectory that minimizes the action integral*

$$A = 2\int_{t_1}^{t_2} T\, dt$$

while at the same time holding total energy H fixed.

Show that this defective version of the principle selects only a few of the trajectories of the spring-mass system CVP 7.

3.23 (**Project**) Find and read Hamilton's original papers wherein he first stated his principle. Trace the evolution of this principle from Newton. Find instances of the numerous misstatements of this principle in various text books.

3.24 Formulate the problem of finding the path of minimum travel time at speed $c = 1$ between any two points on the contour surface $f(w, x, y, z) = 0$.

3.25 Formulate the Lagrangian of the motion of a bead under the downward acceleration g of gravity that is sliding along the parabola $y = x^2$ as the parabola is rotated at constant angular velocity ω about the y-axis.

3.26 (**Project**) Survey the various formulations of the cruise-climb problem (CVP 5). In particular, discuss the planning of the shuttle reentry from orbit.

3.27 (Zenodoros, ca. 180 BC) Formulate the problem of finding the solid of revolution of maximal volume with a given surface area.

3.28 Formulate the problem of finding the solid of revolution of minimal surface area with a given volume.

3.29 A boat must cross a wide river from one dock to another. The river current is known to vary with distance from shore. Formulate the problem of determining the crossing of least time.

Difficulty: The boat makes constant headway v with respect to the water.

3.30 (S. A. Chaplygin) A small airplane is orbiting an airfield at constant airspeed against a steady north wind in time T. Formulate the problem of finding the (closed) orbit of maximal area.

3.31 Formulate the problem of finding the probability distribution of mean 0 and standard deviation 1 of least skewness. Also formulate the problem of least kurtosis.

3.32 A steel beam of uniform width is supported at each end. Formulate the problem of varying the depth of this beam in order to minimize the root-mean-square deflection from the horizontal induced by its own weight.

Chapter 4
The Euler-Lagrange Equation

Our goal in this chapter is to uncover the famous differential equation that must be satisfied when variational integrals are minimized. We then resolve the classic variational problems formulated in Chapter 3 and develop Hamilton's equations predicting the trajectories of conservative mechanical systems. Finally, technical details underlying these famous calculations are examined in more detail.

4.1 One Degree of Freedom

Which path $y = y(x)$ chosen from a prescribed collection C of curves minimizes the cost integral

$$J = \int_a^b L(x, y(x), y'(x))\,dx? \qquad (4.1)$$

[Assume that $L(x, y, z)$ has continuous first partial derivatives everywhere.] This much is clear: If $y = y(x)$ in C achieves the minimum value of the cost J, then any slight perturbation of this path also in C will not decrease J. That is, let $\eta(x)$ be an arbitrary but sufficiently differentiable function on $[a, b]$ with $\eta(a) = 0 = \eta(b)$. Then the cost of using the nearby path $y = y(x) + \epsilon\eta(x)$ is[1]

$$J(\epsilon) = \int_a^b L(x, y(x) + \epsilon\eta(x), y'(x) + \epsilon\eta'(x))\,dx. \qquad (4.2)$$

The cost $J(\epsilon)$ must be at least $J(0)$ for all small ϵ. But since $J(\epsilon)$ is minimum at $\epsilon = 0$, its derivative must vanish at $\epsilon = 0$. The chain rule now makes possible one of the most famous calculations in all of mathematics:

---

[1] Assume the structure of the collection C is such that the perturbed curve $y = y(x) + \epsilon\eta(x)$ remains in C for all small ϵ for each such η.

$$0 = \frac{dJ(\epsilon)}{d\epsilon}\bigg|_{\epsilon=0} = \int_a^b \frac{\partial L(x, y(x) + \epsilon\eta(x), y'(x) + \epsilon\eta'(x))}{\partial \epsilon}\bigg|_{\epsilon=0} dx$$

$$= \int_a^b \left[\frac{\partial L(x, y, y')}{\partial y}\eta + \frac{\partial L(x, y, y')}{\partial y'}\eta'\right] dx. \quad (4.3)$$

By the product rule, the second term of the integrand of (4.3) is

$$\frac{\partial L(x, y, y')}{\partial y'}\eta' = \frac{d}{dx}\left(\frac{\partial L(x, y, y')}{\partial y'}\eta\right) - \left(\frac{d}{dx}\frac{\partial L(x, y, y')}{\partial y'}\right)\eta. \quad (4.4)$$

Replacing the second term in (4.3) with (4.4) and integrating yields

$$J'(0) = 0 = \int_a^b \eta\left[\frac{\partial L}{\partial y} - \frac{d}{dx}\frac{\partial L}{\partial y'}\right] dx + \frac{\partial L(x, y(x), y'(x))}{\partial y'}\eta(x)\bigg|_a^b$$

$$= \int_a^b \eta\left[\frac{\partial L}{\partial y} - \frac{d}{dx}\frac{\partial L}{\partial y'}\right] dx + 0, \quad (4.5)$$

since $\eta(a) = 0 = \eta(b)$—so chosen to ensure that the perturbed path $y = y(x) + \epsilon\eta(x)$ also begins at $(a, y(a))$ and ends at $(b, y(b))$.

Thus from (4.5),

$$\int_a^b \eta\left[\frac{\partial L}{\partial y} - \frac{d}{dx}\frac{\partial L}{\partial y'}\right] dx = 0 \quad (4.6)$$

for arbitrary η. This can happen only when the other factor of the integrand is zero (Exercise 4.1). And that is our grand result, arguably the most useful of all mathematical facts:

Theorem A. (Euler-Lagrange equation) The cost (4.1) is minimized by the path $y = y(x)$ only when

$$\frac{d}{dx}\frac{\partial L}{\partial y'} - \frac{\partial L}{\partial y} = 0. \quad (4.7)$$

Remark A. The preceding computations leading to (4.7) will also hold under the additional constraint that one or both of the endpoint values $y(a)$ and $y(b)$ are specified since our perturbations preserve these values. Moreover, since (4.7) follows from the assumption

4.1 One Degree of Freedom

$J'(0) = 0$, it must hold for merely stationary (critical) paths $y = y(x)$ of J. See (3.9).

Remark B. There is a long tradition of misuse of notation in this subject. In the preceding calculations, we are using the symbols y and y' in three distinct ways: as free variables of the Lagrangian $L(x, y, y')$, as representing a generic curve $y = y(x)$ and its derivative, and as the symbols for the critical curve $y = y(x)$ and its derivative that minimize the cost J.

In the concluding section of this chapter we will take a closer look at the underlying assumptions that make the preceding calculation leading to (4.7) go through. But first let us gain intuition by working through many examples of the application of the Euler-Lagrange equation (4.7) to practical problems.

Example 1. Consider CVP 1 of §3.1, the quest for the shortest path between two points. We are to minimize the arclength

$$J = \int_0^b \sqrt{1 + y'^2} \, dx.$$

By Euler-Lagrange, the shortest continuously differentiable path $y = y(x)$ must satisfy the differential equation

$$0 = \frac{d}{dx} \frac{\partial \sqrt{1 + y'^2}}{\partial y'} - \frac{\partial \sqrt{1 + y'^2}}{\partial y}$$

$$= \frac{d}{dx} \frac{y'}{\sqrt{1 + y'^2}} - 0,$$

and so

$$\frac{y'}{\sqrt{1 + y'^2}} = c. \tag{4.8}$$

Since $c^2 \neq 1$ (why?), we see

$$y'^2 = \frac{c^2}{1 - c^2}, \tag{4.9}$$

that is, y' is constant, and hence $y = y(x)$ is a linear function. The shortest distance between two points is the straight line connecting them.

4.2 Two Special Cases: No y, No x

The Euler-Lagrange equation (4.7) simplifies when the Lagrangian L takes on one of two special forms.

Result A. (no y) When the Lagrangian does not contain y explicitly [i.e., $L = L(x, y')$], the Euler-Lagrange equation (4.7) can be integrated to become the simple relation

$$\frac{\partial L}{\partial y'} = c. \tag{4.10}$$

Proof. Obvious.

The Lagrangians of CVP 1 and 2 have this special form and can be expected to have easy solutions. We have solved CVP 1 previously. Let us now solve CVP 2.

Example 2. To find the graph with least surface of revolution (CVP 2) we must minimize the integral

$$I = \int_a^b x\sqrt{1+y'^2}\, dx. \tag{4.11}$$

See Figure 3.2. By the "no y" result,

$$\frac{\partial L}{\partial y'} = \frac{xy'}{\sqrt{1+y'^2}} = \gamma \tag{4.12}$$

for some constant γ. If $0 \in [a, b]$ or $\gamma = 0$ we obtain a horizontal line and $c = d$. Assume then without loss of generality that $0 < a$ and $c < d$ so the curve is in the position of Figure 3.2. We may then assume that our curve $y = y(x)$ of minimal surface area of revolution is strictly increasing on $[a, b]$ (Exercise 4.3), that is, that $\gamma > 0$. Squaring (4.12) gives

$$(x^2 - \gamma^2)y'^2 = \gamma^2.$$

Since both $y', \gamma > 0$,

$$y = \gamma \int \frac{dx}{\sqrt{x^2 - \gamma^2}} = \gamma\theta + \beta \tag{4.13}$$

4.2 Two Special Cases: No y, No x

by means of the substitution $x = \gamma \cosh \theta$. Therefore our extremal curve is

$$\cosh(\frac{y-\beta}{\gamma}) = \frac{x}{\gamma}. \tag{4.14}$$

We may solve for the values of β, γ using the values for a, b, c, d (Exercise 4.4).

Result B. (no x) In the case that the Lagrangian does not contain x explicitly [i.e., $L = L(y, y')$], the Euler-Lagrange equation (4.7) can be integrated into the form

$$L - y'\frac{\partial L}{\partial y'} = c \tag{4.15}$$

on each interval where the extremizing curve $y = y(x)$ is continuously differentiable.

Proof (under the assumption that y is twice-differentiable; see Exercise 4.53). Note that by the chain rule,

$$\frac{d}{dx}(L - y'\frac{\partial L}{\partial y'}) = \frac{\partial L}{\partial y}y' + \frac{\partial L}{\partial y'}y'' - y''\frac{\partial L}{\partial y'} - y'\frac{d}{dx}\frac{\partial L}{\partial y'}$$

$$= \frac{\partial L}{\partial y}y' - y'\frac{d}{dx}\frac{\partial L}{\partial y'} = 0.$$

Example 3. Consider the catenary (CVP 3) of §3.3, the shape of a hanging string. We search for a continuously differentiable curve $y = y(x)$ that minimizes the integral

$$I = \int_0^a y\sqrt{1+y'^2}\,dx. \tag{4.16}$$

By the "no x" result,

$$y\sqrt{1+y'^2} - \frac{yy'^2}{\sqrt{1+y'^2}} = c, \tag{4.17}$$

(Exercise 4.7). Multiplying through by the radical and squaring gives

$$y^2 = c^2(1+y'^2), \tag{4.18}$$

and so (Exercise 4.9), assuming y even, we obtain the famous result

$$\frac{y}{c} = \cosh \frac{x}{c}. \tag{4.19}$$

But is this valid? We have not considered the constraint of constant arclength. See Exercise 5.38 for a final resolution.

Example 4. The brachistochrone (CVP 4). Let us solve the problem that gave birth to this subject.[2] As we saw in §3.4, our objective is to find the curve $y = y(x)$ that minimizes the integral

$$I = \int_0^b \frac{\sqrt{1+y'^2}}{\sqrt{y}} dx. \tag{4.20}$$

By the "no x" result (4.15), we see that

$$c = \frac{\sqrt{1+y'^2}}{\sqrt{y}} - y'\frac{\partial}{\partial y'}\frac{\sqrt{1+y'^2}}{\sqrt{y}} = \frac{\sqrt{1+y'^2}}{\sqrt{y}} - \frac{y'^2}{\sqrt{y}\sqrt{1+y'^2}}. \tag{4.21}$$

Multiplying (4.21) through by $\sqrt{y}\sqrt{1+y'^2}$ and squaring yields the simple

$$y(1+y'^2) = \gamma^2. \tag{4.22}$$

Make the reasonable assumption that *the optimal path is everywhere sloping downward*. Thus $y = y(x)$ is increasing and $y'(x) \geq 0$. Moreover, from (4.22), $p = y'(x)$ is decreasing and so can be used to parametrize our solution. Thus we have

$$y = \frac{\gamma^2}{1+p^2}, \tag{4.23}$$

where $y = 0$ when $p = \infty$. But then by parts,

$$\frac{x}{\gamma^2} = \frac{1}{\gamma^2}\int \frac{dx}{dy} dy = \int \frac{1}{p} d(\frac{1}{1+p^2}) = \frac{1}{p}\frac{1}{1+p^2} + \int \frac{1}{p^2} \cdot \frac{1}{1+p^2} dp$$

$$= \frac{1}{p(1+p^2)} + \int (\frac{1}{p^2} - \frac{1}{1+p^2}) dp = \frac{1}{p(1+p^2)} - \frac{1}{p} - \text{Arctan } p + C$$

---

[2]This solution is a composite of contributions of Johann Bernoulli, Jakob Bernoulli, G. W. von Leibniz, and Isaac Newton [Katz].

4.3 Multiple Degrees of Freedom

$$= -\frac{p}{1+p^2} - \text{Arctan } p + C.$$

Reparametrizing once again with θ where $p = \tan\theta$, $0 \le \theta \le \pi/2$, we have our curve of least time:

$$x/\gamma^2 = -\sin\theta\cos\theta - \theta + C \qquad (4.24a)$$

and

$$y/\gamma^2 = \cos^2\theta. \qquad (4.24b)$$

Since $x = 0 = y$ at $\theta = \pi/2$,

$$C = \frac{\pi}{2}. \qquad (4.24c)$$

For example, taking $\gamma = 1$ to obtain the terminal point $(\pi/2, 1)$ at $\theta = 0$, the *Mathematica* script

`ParametricPlot[-Sin[t]*Cos[t] - t + Pi/2,Cos[t]*Cos[t],{t,0,Pi/2}]`

yields the shape shown in Figure 4.1.

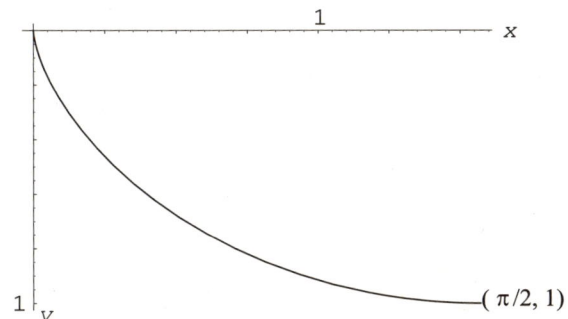

Figure 4.1 The cycloid, the path of least time of descent.

This shape is known as a *cycloid*, since from (4.24), with $\phi = \pi - 2\theta$, $0 \le \phi \le \pi$,

$$2x/\gamma^2 = \phi - \sin\phi \qquad \text{and} \qquad 2y/\gamma^2 = 1 - \cos\phi, \qquad (4.25)$$

which is the curve traced by a red dot on a bicycle tire of radius $r = \gamma^2/2$ as it rolls rightward in Figure 4.1 along the x-axis. But (4.25) reveals the unexpected consequence that not all desired destinations (b, d) are obtainable with these everywhere downward sloping paths! To reach some destinations, we must momentarily dip below the final level by continuing the path for a portion of $\pi < \phi < 2\pi$. See Exercises 4.41–4.47.

4.3 Multiple Degrees of Freedom

Suppose we have a conservative mechanical system whose configuration is determined by n independent *generalized coordinates* $q = (q_1, q_2, \ldots, q_n)$, which has kinetic energy T, and potential energy V. All trajectories actually taken by this system must by Hamilton's principle (§3.7) be critical curves for the integral

$$J = \int_{t_1}^{t_2} L(t, q, \dot{q}) \, dt, \tag{4.26}$$

where the Lagrangian $L = T - V$. By additively perturbing the critical path $q = q(t)$ by a vector $\eta = \eta(t)$, where $\eta(t_1) = 0 = \eta(t_2)$, we obtain a new cost

$$J(\epsilon) = \int_{t_1}^{t_2} L(t, q(t) + \epsilon \eta(t), \dot{q}(t) + \epsilon \dot{\eta}(t)) \, dt, \tag{4.27}$$

giving by the chain rule that

$$\frac{dJ(0)}{d\epsilon} = 0 = \int_{t_1}^{t_2} \sum_{k=1}^{n} [\eta_k \frac{\partial L}{\partial q_k} + \dot{\eta}_k \frac{\partial L}{\partial \dot{q}_k}] \, dt \tag{4.28}$$

$$= \text{(Exercise 4.10)} = \int_{t_1}^{t_2} \sum_{k=1}^{n} \eta_k [\frac{\partial L}{\partial q_k} - \frac{d}{dt}\frac{\partial L}{\partial \dot{q}_k}] \, dt. \tag{4.29}$$

Because we are at liberty to choose all but one $\eta_k(x) \equiv 0$, we again obtain from Exercise 4.1 the fundamental law of mechanics.

Theorem B. (Lagrange equations) A conservative mechanical system can evolve only along trajectories $q = q(t)$ that satisfy

$$\frac{d}{dt}\frac{\partial L}{\partial \dot{q}_k} - \frac{\partial L}{\partial q_k} = 0, \qquad k = 1, 2, \ldots, n. \tag{4.30}$$

Example 5. To practice the use of (4.30), let us obtain the ODE that governs the motion of the one degree-of-freedom mass-spring system CVP 7 of Figure 3.7. Since $L = m\dot{q}^2/2 - kq^2/2$,

$$\frac{d}{dt}\frac{\partial L}{\partial \dot{q}} - \frac{\partial L}{\partial q} = 0 = m\ddot{q} + kq,$$

4.3 Multiple Degrees of Freedom

that is,
$$m\ddot{q} = -kq. \tag{4.31}$$

Example 6. An extensible planar pendulum (CVP 11). Consider a planar pendulum with a telescoping arm of length r as shown in Figure 4.2. The telescoping action is restrained by an internal spring of constant k so that at rest (at bottom dead center), $r = r_0$.

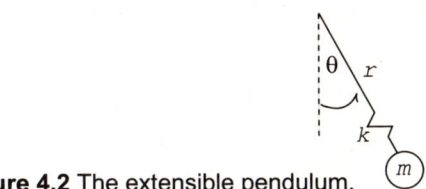

Figure 4.2 The extensible pendulum.

The kinetic energy of this system is the sum of its radial and tangential kinetic energies (since these respective velocities are orthogonal):
$$T = \frac{m\dot{r}^2}{2} + \frac{mr^2\dot{\theta}^2}{2}, \tag{4.32a}$$
while its potential energy is the work done stretching the spring plus lifting the mass:
$$V = k(r - r_0)^2/2 + mg(r_0 - r\cos\theta). \tag{4.32b}$$

The Lagrange equations (4.30) become (Exercise 4.21)
$$m\ddot{r} - mr\dot{\theta}^2 + k(r - r_0) - mg\cos\theta = 0, \tag{4.33a}$$
$$2mr\dot{r}\dot{\theta} + mr^2\ddot{\theta} - mgr\sin\theta = 0. \tag{4.33b}$$

Result C. (no t) In the special case that the Lagrangian is time invariant, that is, $L = L(q, \dot{q})$,
$$L - \sum_{k=1}^{n} \dot{q}_k \frac{\partial L}{\partial \dot{q}_k} = c \tag{4.34}$$

(Exercise 4.23). That is, the left hand side of (4.34) is an *integral of motion*, constant along any given mechanical trajectory.

Definition. The negative of the left-hand side of (4.34),

$$H = \sum_{k=1}^{n} \dot{q}_k \frac{\partial L}{\partial \dot{q}_k} - L, \tag{4.35}$$

is called the *Hamiltonian* of the system. It is constant along each mechanical trajectory $(q(t), \dot{q}(t))$.

4.4 The Hamiltonian

Assume that the Lagrangian of a mechanical system is time invariant [i.e., $L = L(q, \dot{q})$], that $V = V(q)$, and that T is quadratic in $\dot{q}$ [i.e., $T(q, \lambda\dot{q}) = \lambda^2 T(q, \dot{q})$, where $q, \dot{q}$ are the vector variables $q = (q_1, q_2, \ldots, q_n)^\top$ and $\dot{q} = (\dot{q}_1, \dot{q}_2, \ldots, \dot{q}_n)^\top$]. Last, assume kinetic energy is positive definite in $\dot{q}$ [i.e., $T(q, \dot{q}) > 0$ whenever $\dot{q} \neq 0$].

It then follows directly from Euler's formula (Exercise 4.24) that

$$2T = \sum_{k=1}^{n} \dot{q}_k \frac{\partial T}{\partial \dot{q}_k} = \sum_{k=1}^{n} \dot{q}_k \frac{\partial L}{\partial \dot{q}_k} = H + L, \tag{4.36}$$

and hence *the Hamiltonian (4.35) is total energy*, that is,

$$H = T + V. \tag{4.37}$$

Introduce the new variables of *generalized momenta*:

$$p_k = \frac{\partial L}{\partial \dot{q}_k} = \frac{\partial T}{\partial \dot{q}_k}. \tag{4.38}$$

The Hamiltonian (4.35) becomes

$$H = \sum_{k=1}^{n} p_k \dot{q}_k - L. \tag{4.39}$$

Lemma. When $\dot{q}, p$ are written as column vectors, they are related by the matrix rule

$$p = \left[\frac{\partial^2 L}{\partial \dot{q}_i \partial \dot{q}_j}\right] \dot{q}. \tag{4.40}$$

Although the entries of the matrix may depend upon q, they are all free of dependence on $\dot{q}$.

4.4 The Hamiltonian

Proof. From (4.36) and the product rule,

$$2p_i = 2\frac{\partial T}{\partial \dot{q}_i} = p_i + \sum_{j=1}^{n} \dot{q}_j \frac{\partial^2 T}{\partial \dot{q}_i \partial \dot{q}_j}.$$

This can also be seen from the fact that $\partial T/\partial \dot{q}_i$ is homogeneous of degree 1 in $\dot{q}$ since it is the partial of a function of degree 2.

Carrying homogeneity one step further, the entries of the matrix of (4.40) are second partials and hence homogeneous of degree 0 in $\dot{q}$, that is,

$$\frac{\partial^2 L}{\partial \dot{q}_i \dot{q}_j} = F^{i,j}(q, \dot{q}),$$

where

$$F^{i,j}(q, \lambda \dot{q}) = F^{i,j}(q, \dot{q}) = F^{i,j}(q, 0).$$

Corollary. The matrix of (4.40) relating the variables $\dot{q}$ to p is invertible.

Proof. For $\dot{q} \neq 0$,

$$\dot{q}^T \left[\frac{\partial^2 L}{\partial \dot{q}_i \partial \dot{q}_j} \right] \dot{q} = \dot{q}^T p = 2T > 0.$$

Remark. The lemma and its corollary mean that it is possible to replace the two independent *configuration* variables $q, \dot{q}$ by the so called *canonical* variables q, p. In particular, the Hamiltonian can be rewritten (in theory) in the form $H = H(q, p)$ with $\dot{q}$ not appearing explicitly.

More carefully, since the quantities $\dot{q}$ and p are related by an invertible matrix relation $p = A(q)\dot{q}$, once q is chosen and fixed, we may choose $\dot{q}$ freely thus determining the value for p, or conversely, we may choose p thus determining the value of $\dot{q}$.

Theorem D. (Hamilton's equations) A time-invariant mechanical system with kinetic energy T that is quadratic and positive definite in $\dot{q}$ with potential energy $V = V(q)$ must follow trajectories determined by

$$\dot{q}_k = \frac{\partial H}{\partial p_k} \quad \text{and} \quad \dot{p}_k = -\frac{\partial H}{\partial q_k}. \qquad (4.41)$$

Proof. To obtain this fundamental result, we must be very clear about the meaning of the partial derivatives taken. There are ambiguities: When differentiating with respect to one variable, what other variables must be held fixed? Does $\partial L/\partial q_k$ mean we hold the p-variables fixed or the $\dot{q}$-variables fixed? To this end we borrow notation common in thermodynamics: For example,

$$\left(\frac{\partial F}{\partial q_k}\right)_p$$

will denote the partial of F with respect to q_k holding all p_j and all other q_j fixed.

From (4.39) together with the product and chain rules comes

$$\frac{\partial H}{\partial p_k} = \left(\frac{\partial H}{\partial p_k}\right)_q = \dot{q}_k + \sum_{j=1}^n p_j \left(\frac{\partial \dot{q}_j}{\partial p_k}\right)_q - \left(\frac{\partial L}{\partial p_k}\right)_q$$

$$= \dot{q}_k + \sum_{j=1}^n p_j \left(\frac{\partial \dot{q}_j}{\partial p_k}\right)_q - \sum_{j=1}^n \left(\frac{\partial L}{\partial q_j}\right)_{\dot{q}} \left(\frac{\partial q_j}{\partial p_k}\right)_q - \sum_{j=1}^n \left(\frac{\partial L}{\partial \dot{q}_j}\right)_q \left(\frac{\partial \dot{q}_j}{\partial p_k}\right)_q$$

$$= \dot{q}_k + \sum_{j=1}^n p_j \left(\frac{\partial \dot{q}_j}{\partial p_k}\right)_q - \sum_{j=1}^n p_j \left(\frac{\partial \dot{q}_j}{\partial p_k}\right)_q = \dot{q}_k.$$

Again from (4.39) via the chain rule and the Lagrange equations,

$$\frac{\partial H}{\partial q_k} = \left(\frac{\partial H}{\partial q_k}\right)_p = \sum_{j=1}^n p_j \left(\frac{\partial \dot{q}_j}{\partial q_k}\right)_p - \left(\frac{\partial L}{\partial q_k}\right)_p$$

$$= \sum_{j=1}^n p_j \left(\frac{\partial \dot{q}_j}{\partial q_k}\right)_p - \sum_{j=1}^n \left(\frac{\partial L}{\partial q_j}\right)_{\dot{q}} \left(\frac{\partial q_j}{\partial q_k}\right)_p - \sum_{j=1}^n \left(\frac{\partial L}{\partial \dot{q}_j}\right)_q \left(\frac{\partial \dot{q}_j}{\partial q_k}\right)_p$$

$$= \sum_{j=1}^n p_j \left(\frac{\partial \dot{q}_j}{\partial q_k}\right)_p - \left(\frac{\partial L}{\partial q_k}\right)_{\dot{q}} - \sum_{j=1}^n p_j \left(\frac{\partial \dot{q}_j}{\partial q_k}\right)_p$$

$$= -\left(\frac{\partial L}{\partial q_k}\right)_{\dot{q}} = -\frac{d}{dt}\left(\frac{\partial L}{\partial \dot{q}_k}\right)_q = -\frac{d}{dt}p_k = -\dot{p}_k.$$

4.4 The Hamiltonian

Example 6. (reprise) Let us rewrite the extensible planar pendulum (CVP 11) of Figure 4.2 in Hamilton's form. Recall that

$$T = \frac{m\dot{r}^2}{2} + \frac{mr^2\dot{\theta}^2}{2}$$

and

$$V = k(r - r_0)^2/2 + mg(r_0 - r\cos\theta).$$

Setting $q_1 = \theta$ and $q_2 = r$, the kinetic energy becomes

$$T = \frac{m\dot{q}_2^2}{2} + \frac{mq_2^2 \dot{q}_1{}^2}{2}, \tag{4.42}$$

and so

$$p_1 = \frac{\partial T}{\partial \dot{q}_1} = mq_2^2 \dot{q}_1 \quad \text{and} \quad p_2 = \frac{\partial T}{\partial \dot{q}_2} = m\dot{q}_2, \tag{4.43}$$

a form guaranteed by (4.40). We can rewrite kinetic energy using these generalized momenta p_k as

$$T = \frac{p_1^2}{2mq_2^2} + \frac{p_2^2}{2m}. \tag{4.44}$$

The Hamiltonian is therefore

$$H(q,p) = \frac{p_1^2}{2mq_2^2} + \frac{p_2^2}{2m} + \frac{k(q_2 - r_0)^2}{2} + mg(r_0 - q_2 \cos q_1). \tag{4.45}$$

Hamilton's equations are then (Exercise 4.28)

$$\dot{q}_1 = \frac{\partial H}{\partial p_1} = \frac{p_1}{mq_2^2}, \tag{4.46a}$$

$$\dot{q}_2 = \frac{\partial H}{\partial p_2} = \frac{p_2}{m}, \tag{4.46b}$$

$$\dot{p}_1 = -\frac{\partial H}{\partial q_1} = -mgq_2 \sin q_1, \tag{4.46c}$$

$$\dot{p}_2 = -\frac{\partial H}{\partial q_2} = \frac{p_1^2}{mq_2^3} - k(q_2 - r_0) + mg \cos q_1. \tag{4.46d}$$

We have exchanged the Lagrange system of two coupled second-order ODEs for Hamilton's system of four coupled first-order equations.

4.5 A Closer Look

What exactly are the hypotheses under which the proof of the Euler-Lagrange equation (4.7) is valid? Here are the steps once again:

Suppose the path $y = y(x)$ minimizes the integral

$$J = \int_a^b L(x, y(x), y'(x)) \, dx. \tag{4.47}$$

Thus if $\eta(x)$ is any continuously differentiable function on $[a, b]$ with $\eta(a) = 0 = \eta(b)$, the cost of using the nearby path $y = y(x) + \epsilon\eta(x)$ is

$$J(\epsilon) = \int_a^b L(x, y(x) + \epsilon\eta(x), y'(x) + \epsilon\eta'(x)) \, dx. \tag{4.48}$$

But since $J(\epsilon)$ is minimum at $\epsilon = 0$, its derivative must vanish at $\epsilon = 0$. Hence by the chain rule,

$$0 = \left.\frac{dJ(\epsilon)}{d\epsilon}\right|_{\epsilon=0} = \int_a^b \left.\frac{\partial L(x, y(x) + \epsilon\eta(x), y'(x) + \epsilon\eta'(x))}{\partial \epsilon}\right|_{\epsilon=0} dx$$

$$= \int_a^b [L_y(x, y, y')\eta + L_{y'}(x, y, y')\eta'] \, dx. \tag{4.49}$$

By integrating the second term by parts we obtain

$$0 = \int_a^b [L_y - \frac{d}{dx}L_{y'}]\eta \, dx + L_{y'}(x, y(x), y'(x))\eta(x)\big|_a^b, \tag{4.50}$$

and therefore, since $\eta(a) = 0 = \eta(b)$,

$$\int_a^b [L_y - \frac{d}{dx}L_{y'}]\eta \, dx = 0 \tag{4.51}$$

for arbitrary η. This can happen only when the first factor of the integrand is zero and hence the Euler-Lagrange equation obtains.

Critique. We may differentiate past the integral sign and apply the chain rule to obtain (4.49) by assuming the continuity of the first partials of the Lagrangian L on the (open connected) domain likely to contain minimizing curves and by assuming y' and η' are bounded (Exercise 4.2). But the integration by parts in the next line (4.50)

4.5 A Closer Look

appears to require the differentiability of $L_{y'}$ and the integrability of the resulting derivative; if y' appears explicitly in the result of this partial, y'' must apparently exist and be integrable. But this is far too much to require.

What is actually needed is only that $y = y(x)$ be *absolutely continuous with a bounded derivative,* a stronger requirement than continuity but weaker than requiring that y be piecewise continuously differentiable. A function f is absolutely continuous if it arises as an integral:

$$f(x) = f(a) + \int_a^x v(t)\, dt. \tag{4.52}$$

For then f is differentiable almost everywhere with the integrable derivative v [Rudin].

But how can the derivation of the Euler-Lagrange equation (a second-order ODE in y) be valid when the second derivative y'' is not assumed to exist necessarily? The answer lies in taking another fork in the proof. From (4.49) we have by integration by parts (Exercise 4.34) of the *first* term that

$$0 = \int_a^b [L_y(x,y,y')\eta + L_{y'}(x,y,y')\eta']\, dx \tag{4.53}$$

$$= \eta(x) \int_a^x L_y(t,y,y')dt \Big|_a^b - \int_a^b [\int_a^x L_y(t,y,y')dt - L_{y'}(x,y,y')]\eta'\, dx$$

$$= \int_a^b [\int_a^x L_y(t,y,y')\, dt - L_{y'}(x,y,y')]\eta'\, dx. \tag{4.54}$$

But then by the du Bois-Reymond Lemma (Exercise 4.35)—provided $L_{y'}$ is continuous along the critical curve (see Exercise 11.18)—we obtain an integral form of the Euler-Lagrange equation:

$$\int_a^x L_y(t,y,y')\, dt - L_{y'}(x,y,y') = c, \tag{4.55}$$

sometimes called the *fundamental lemma of the calculus of variations* [Pierre]. Paths $y = y(x)$ satisfying (4.55) are called *extremals* of the cost J of (4.1). We may now bootstrap from (4.55) to see that

$$g(x) = L_{y'}(x,y,y')$$

is absolutely continuous with a bounded derivative, which justifies the integration by parts used to obtain (4.50). Thus stationary (critical) curves induce additional regularity.

This famous proof scheme (4.47)–(4.51) applies with little modification to a wide class of generalizations, as we shall see repeatedly in subsequent chapters. Consult [McShane] for a historical sketch of the origin of these ideas and for his expert and sometimes unflattering critique of their implications.

EXERCISES

4.1 For f continuous, show that if

$$\int_a^b \eta(x) f(x)\, dx = 0$$

for all infinitely differentiable η with $\eta(a) = 0 = \eta(b)$, then $f(x) \equiv 0$ on $[a, b]$. (This result also holds when f is merely integrable—see Exercise 4.52.)

Hint: Look at an interval where f is positive.

4.2* Verify (4.3) obtains given that L has continuous first partials and that y' and η' are bounded on $[a, b]$.

4.3 Show that in CVP 2, the curve of least revolved surface area is given by either a strictly increasing or decreasing function $y(x)$. Give two arguments—one geometric, the other using (4.12).

4.4 Attempt to find a formulae for β and γ of (4.14) in terms of a, b, c, d.

4.5 Using numerical methods with choices $a = c = 1$ and $b = d = 2$, graph the solution (4.14).

4.6 To obtain appreciation of the "no x" result, attempt to solve the catenary (CVP 3) of §3.3 using the Euler-Lagrange equation directly.

4.7 Deduce (4.17) from (4.16).

Exercises

4.8 What are the untoward consequences of placing the coordinate system in Figure 3.3 so that the bottom of the curve is at $(0,0)$? Examine (4.17). Explain the apparent contradiction.

4.9 Establish (4.18) and (4.19).

4.10 Deduce (4.29) from (4.28) via the product rule.

4.11 Use the "no x" result to obtain an ODE governing the motion of the mass-spring system of Figure 3.7.

Answer: $m\dot{q}^2 + kq^2 = c$.

4.12 Find the ODE that governs the motion of the planar pendulum of Figure 3.8 via the Euler-Lagrange equation.

Answer: $a\ddot{\theta} = -g\sin\theta$.

4.13 Find the ODE that governs the motion of the double planar pendulum of Exercise 3.12 via the Euler-Lagrange equation.

Answer: $(ma^2 + MA^2)\ddot{\theta} = -(MgA - mga)\sin\theta$.

4.14 Find the system of ODEs that governs the motion of the multiple mass spring system of Exercise 3.7 via the Lagrange equations (4.30).

Answer: $m_i\ddot{q}_i = k_{i-1}q_{i-1} - (k_{i-1} + k_i)q_i + k_i q_{i+1}$.

4.15 Find the system of ODEs that governs the motion of the articulated pendulum of Figure 3.9.

4.16 State and prove a "no y" result [i.e., what does the Euler-Lagrange equation (4.7) reduce to in the case $L = L(x, y)$]?

4.17 A swimmer wishes to reach the opposite bank in the shortest time against a steady current. What path should she take?

4.18 Find the path $y = y(x)$ through the points $(0, 1)$ and $(1, 1)$ that minimizes the integral

$$J = \int_0^1 (y'^2 + 2y' + y^2)\, dx.$$

4.19 Find the path $y = y(x)$ through the points $(0,1)$ and $(1,1)$ that minimizes the integral

$$J = \int_0^1 \sqrt{y'}\sqrt{1+y'^2}\, dx.$$

4.20 Find the path $y = y(x)$ through the points $(0,1)$ and $(1,1)$ that minimizes the integral

$$J = \int_0^1 \left(\sqrt{1+y'^2} - \sqrt{1-y^2}\right) dx.$$

4.21 Carefully verify the formulae (4.33) for the extensible pendulum.

4.22 Solve for Newton's shape of minimum resistance of §3.6.

Outline: Using the "no y" result $xy' = -C(1+y'^2)^2$, show that $p = -y' > 0$ is monotonic as a function of x for $p > 1/\sqrt{3}$ and so can be used to parameterize both x and y, namely, $x = C(1+p^2)^2/p$ and

$$-y/C = \int p\, dx/C = \cdots = D + \log p - p^2 - 3p^4/4.$$

4.23 Deduce the "no t" formula (4.34) for multi-degree-of-freedom Lagrangians from (4.29).

4.24 Prove *Euler's formula:* If F is *homogeneous* of degree k, that is,

$$F(\lambda x_1, \lambda x_2, \ldots, \lambda x_n) = \lambda^k F(x_1, x_2, \ldots, x_n),$$

then

$$\sum_{j=1}^n x_j \frac{\partial F(x)}{\partial x_j} = kF(x).$$

4.25 Write down Hamilton's equations for the mass-spring system CVP 7.

Answer: $\dot{q} = p/m$ and $\dot{p} = -kq$.

4.26 Write down Hamilton's equations for the planar pendulum CVP 8.

Exercises

4.27 Write the Hamiltonian of the articulated pendulum CVP 9 of §3.7 in the form $H = H(q,p)$.

4.28 Verify the Hamilton equations (4.46) for the extensible pendulum.

4.29 (**Project**) Redo §4.4 for Lagrangians $L = L(t,q,\dot{q})$ that are not necessarily time-invariant.

4.30 Find the Lagrange equations for a spherical pendulum.

4.31 Find Hamilton's equations for a spherical pendulum.

4.32 Show that if $q = q(t)$ minimizes the cost

$$J = \int_a^b L(t,q,\dot{q},\ddot{q})\, dt,$$

then

$$\frac{\partial L}{\partial q_k} - \frac{d}{dt}\frac{\partial L}{\partial \dot{q}_k} + \frac{d^2}{dt^2}\frac{\partial L}{\partial \ddot{q}_k} = 0.$$

4.33 Show that any minimizing curve of the cost

$$J = \phi(y(b)) - \psi(y(a)) + \int_a^b L(x,y,y')\, dx$$

must satisfy the Euler-Lagrange equations (4.7).

4.34 Prove *strong integration by parts:* Suppose both $u(x), v(x)$ are integrable. Set

$$U(x) = \int_a^x u(t)\, dt \quad \text{and} \quad V(x) = \int_a^x v(t)\, dt.$$

Then

$$\int_a^b U(x)v(x)\, dx = U(x)V(x)\Big|_a^b - \int_a^b u(x)V(x)\, dx.$$

4.35 (**Du Bois-Reymond**) Suppose F is continuous on $[a,b]$. Prove that if for every infinitely differentiable $\eta = \eta(x)$ with $\eta(a) = 0 = \eta(b)$ we have

$$\int_a^b \eta'(x)F(x)\, dx = 0,$$

then F is a constant function. Compare with Exercise 11.18.

Outline: If F is of mixed sign, find an η that increases on an subinterval of one inverval where $F(x)$ is positive, decreases on a subinterval of one interval where $F(x)$ is negative, and is otherwise constant. This gives a positive integral of the product $\eta' F$. Apply to vertical translates of the graph of F.

4.36 (Transversality) Suppose one is free to search among curves $y = y(x)$ to minimize the cost
$$J = \int_a^b L(x, y, y') \, dx$$
without the burden of satisfying an *end condition* at $x = b$, that is, the value of $y(b)$ can be arbitrary. Prove that not only does the Euler-Lagrange equation (4.7) hold, but in addition we have the *natural boundary condition*
$$L_{y'}(b, y(b), y'(b)) = 0.$$
Hint: Relax the condition $\eta(b) = 0$ in the proof of §4.1.

4.37 (Transversality) Among all piecewise continuously differentiable curves, suppose $q = q(t)$ minimizes the cost
$$J = \int_a^b L(t, q, \dot{q}) \, dt$$
and satisfies the initial condition $q(a) = q_0$ yet the ith target coordinate $q_i(b)$ is unspecified. Prove that we obtain the *natural boundary condition*
$$L_{\dot{q}_i}(b, q(b), \dot{q}(b)) = 0.$$

4.38 Show that the cost
$$J = \int_{-1}^1 x^2 y'^2 \, dx, \quad y(-1) = -1, \ y(1) = 1,$$
has no minimum value among twice continuously differentiable curves $y = y(x)$.

4.39 A bead of mass m is free to slide along the x-axis restrained only by a spring whose other end is connected at $(0, 1)$. The spring has spring constant k and natural length 1. Find the ODE governing the motion of the mass.

Exercises

4.40 Galileo estimated the acceleration g of gravity by rolling solid balls down an inclined plane at extreme inclinations. Show that because of rotational inertia, his estimates of g were doomed to be $5g/7$. He *was* able to demonstrate that velocity increases proportionally with time.

4.41 Show that in the cycloid solution (4.25) to the brachistochrone, $\phi = T\sqrt{2g}/\gamma$, where T is the time the sliding particle has been in motion.

4.42 Show that the time T required to slide down a frictionless ramp of length L inclined at angle α is $T = \sqrt{2L/g \sin \alpha}$.

4.43 Using actual physical constants, determine the length of time T to slide down the cycloid path of minimum time from a height of 100 ft to a spot 100 ft horizontally distant. Compare your result to the time to slide down an inclined plane. (You may use the formula for T from Exercise 4.45 without proof.)

4.44 The downward sloping cycloid path (4.25), displayed as the solution to the brachistochrone problem (CVP 4) in §4.2, is incapable of reaching all destinations (b, d) of Figure 3.4. Prove that regardless of the choice of the constant γ in (4.25), no portion of the (everywhere downward sloping) path can reach points (b, d) above the radial line $y = 2x/\pi$ [i.e., points (b, d) with $d/b < 2/\pi$]. Show, however, that it is possible to meet any vertical line $x = b$ with terminal velocity that is horizontal.

4.45 (**Brachistochrone**) Prove that all final destinations (b, d) of Figure 3.4 with $b, d > 0$ are reachable in least time T along a unique cycloid $2x/\gamma^2 = \phi - \sin\phi$, $2y/\gamma^2 = 1 - \cos\phi$, $0 < \phi < 2\pi$. Show also that this minimal time is given by $T = \gamma\phi/\sqrt{2g}$, where ϕ is the unique solution to

$$\frac{d}{b} = \frac{1 - \cos\phi}{\phi - \sin\phi}, \quad 0 < \phi < 2\pi.$$

4.46 In the formulation of the brachistochrone problem in §2.4, parametrize arclength with y to obtain the cost

$$J = \int_0^d \frac{\sqrt{1 + (dx/dy)^2}}{\sqrt{2gy}} \, dy.$$

In view of this Lagrangian being free of the now dependent variable x, can Result A of §4.2 lead to a simpler solution to the problem?

4.47 How can we be sure that the cycloid (4.25) is the path of *least* time? Could it be merely stationary? Give an intuitive, geometric argument.

4.48 Show that the "no x" result (4.15) may introduce extraneous solutions that do not satisfy the Euler-Lagrange equation (4.7).

Outline. In Exercise 4.18, $y = 1$ satisfies (4.15) but not (4.7).

4.49 Show that not every trajectory of constant energy (4.34) satisfies the Lagrange equations (4.30). Try the system $T = \dot{q}^2/2$ and $V = q - t$. Are there time-invariant examples?

4.50 From the proof of (4.41), it is clear that any trajectory satisfying Lagrange's equations (4.30) must satisfy Hamilton's equations (4.41). But can there be motions satisfying Hamilton's equations that do not satisfy the Lagrange equations[3]?

4.51 For that matter, will each solution of the Lagrange equations (4.30) actually occur as an actual mechanical motion? Is it true (as is often said about Maxwell's equations) that *whatever the mathematics allows, occurs?*

4.52* Deduce that $f = 0$ a.e. in Exercise 4.1 for f merely integrable from the du Bois-Reymond lemma of Exercise 4.35.

Hint: Take
$$F(x) = \int_a^x f(t)\, dt.$$

4.53 Prove a generalization of the "no x" result (4.15): On each open interval I where an extremal $y = y(x)$ of the twice continuously differentiable Lagrangian L is itself twice differentiable, there exists a constant c_I so that
$$L(x, y, y') - y' L_{y'}(x, y, y') - \int_a^x L_x\, dt = c_I.$$

Hint: Employing the integral form of the Euler-Lagrange equation (4.55) for $L_{y'}$, differentiate the left-hand side.

---

[3] and hence cannot occur as actual mechanical motions.

Chapter 5
Constrained Problems

Many important physical optimization problems have side conditions (constraints). The Euler-Lagrange equation is then applied to a new Lagrangian augmented by the constraint. We begin with the canonical example of Dido's isoperimetric problem, then state and solve several other representative problems. We then extend Hamilton's principle to constrained mechanical problems.

5.1 Dido's Problem (CVP 10)

We would like to maximize the area enclosed by a curve $y = y(x)$ of a given arclength γ that uses a given portion $[a, b]$ of the x-axis as part of the boundary. That is, we must maximize the integral

$$J = \int_a^b y\, dx \tag{5.1a}$$

subject to the integral constraint

$$C = \int_a^b \sqrt{1 + y'^2}\, dx = \gamma. \tag{5.1b}$$

Such problems are called *isoperimetric problems* because the constraint fixes the perimeter. It is nowadays common to call any problem with an integral constraint an isoperimetric problem.

The usual Euler-Lagrange equation (4.7) is no longer applicable since the candidate curves $y = y(x)$ are further constrained by (5.1b). An attempt to apply the proof of (4.7) in this situation is doomed since an arbitrary perturbation of $y(x)$ may in fact obtain a more favorable cost J but not satisfy the constraint. The key is to try *two*-parameter perturbations

$$y_1(x) = y(x) + \epsilon \eta(x) + \delta \zeta(x). \tag{5.2}$$

5.2 Statement of the Problem

Our goal is to solve problems of the following type:

Find the curve $y = y(x)$ that minimizes (maximizes) the cost

$$J = \int_a^b L(x, y, y') \, dx \tag{5.3a}$$

while satisfying the integral constraint

$$C = \int_a^b M(x, y, y') \, dx = c_0, \tag{5.3b}$$

and possible end conditions $y(a) = c$, $y(b) = d$. We restrict our search to curves $y = y(x)$ belonging to the class of all absolutely continuous curves with bounded derivatives, or more often to the class of piecewise continuously differentiable curves, or occasionally to an even more restrictive class of curves.

To attack such problems we will need one of the workhorses of mathematics.

5.3 The Inverse Function Theorem

Theorem A. Suppose

$$F(x) = (f_1(x), f_2(x), \ldots, f_n(x)) \tag{5.4}$$

is a vector-valued function of the vector variable $x = (x_1, x_2, \ldots, x_n)$. Suppose that F is continuously differentiable on a neighborhood of x^0 and that its *Jacobian* at $x = x^0$ is nonzero, that is,

$$\det \left[\frac{\partial f_i}{\partial x_j} \right]_{x^0} \neq 0. \tag{5.5}$$

Then there exists an open neighborhood U of x^0 and an open neighborhood V of $y^0 = F(x^0)$ where

$$F : U \to V \tag{5.6}$$

is bijective with a continuously differentiable inverse function $G = F^{-1}$.

5.4 The Euler-Lagrange Equation for Constrained Problems

A proof of this result is found in Appendix A. We now apply this powerful tool to our constrained minimization problem (5.3).

5.4 The Euler-Lagrange Equation for Constrained Problems

Consider two smooth curves $\eta = \eta(x)$ and $\zeta = \zeta(x)$ on the interval $[a, b]$, both vanishing at both endpoints. Define the perturbed cost and perturbed constraint as

$$J(\epsilon, \delta) = \int_a^b L(x, y(x) + \epsilon\eta(x) + \delta\zeta(x), y'(x) + \epsilon\eta'(x) + \delta\zeta'(x)) \, dx \quad (5.7a)$$

and

$$C(\epsilon, \delta) = \int_a^b M(x, y(x) + \epsilon\eta(x) + \delta\zeta(x), y'(x) + \epsilon\eta'(x) + \delta\zeta'(x)) \, dx. \quad (5.7b)$$

We now repeat the steps of the proof of the theorem on Lagrange multipliers of §1.4 for this special case $J + \lambda C$.

Let $J(0, 0) = m_0$ and $C(0, 0) = c_0$, where m_0 is the assumed minimum cost achieved by the curve $y = y(x)$.

If the Jacobian of the mapping

$$F(\epsilon, \delta) = (J(\epsilon, \delta), C(\epsilon, \delta)) \quad (5.8)$$

does not vanish at $(0, 0)$, we can by the inverse function theorem find (ϵ, δ) close to $(0, 0)$ with

$$J(\epsilon, \delta) < m_0 \quad \text{and} \quad C(\epsilon, \delta) = c_0. \quad (5.9)$$

See Figure 5.1.

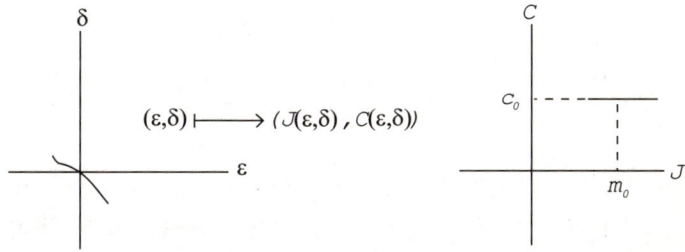

Figure 5.1 If the mapping has nonzero Jacobian at *(0,0)*, then we can find paths with costs *J(ε,δ)* smaller than the optimal value m_0 yet that satisfy the constraint *C(ε,δ) = c_0*.

But this is incompatible with the assumed minimality of y. So the indicated Jacobian must vanish, that is,

$$\det \begin{bmatrix} J_\epsilon & J_\delta \\ C_\epsilon & C_\delta \end{bmatrix}_{(0,0)} = 0 \tag{5.10}$$

for all η and ζ. But a *determinant vanishes if and only if the matrix is singular*, i.e., in this 2×2 case, *one row is a multiple of the other*.

Note from (5.7b) that $C_\epsilon(0,0)$ depends only upon the choice made for η, not ζ. There are now two cases: Either $C_\epsilon(0,0) = 0$ for all η or $C_\epsilon(0,0) \neq 0$ for some η. If the former holds, then the constraint M satisfies the usual Euler-Lagrange equation (4.7) by Exercise 5.1. If the latter holds for some η_0, then for some scalar λ,

$$(J_\epsilon(0,0), J_\delta(0,0)) + \lambda(C_\epsilon(0,0), C_\delta(0,0)) = (0,0) \tag{5.11}$$

for this η_0 and for all ζ. Thus writing out the second component of (5.11) and integrating by parts as in the derivation of the Euler-Lagrange equations in §4.1 yields

$$\int_a^b \zeta[\frac{\partial L}{\partial y} - \frac{d}{dx}\frac{\partial L}{\partial y'} + \lambda \frac{\partial M}{\partial y} - \lambda \frac{d}{dx}\frac{\partial M}{\partial y'}] \, dx = 0, \tag{5.12}$$

for all ζ. But then, setting

$$N(x, y, y') = L(x, y, y') + \lambda M(x, y, y'), \tag{5.13}$$

we see that

$$0 = \int_a^b \zeta[\frac{\partial N}{\partial y} - \frac{d}{dx}\frac{\partial N}{\partial y'}] \, dx \tag{5.14}$$

for all ζ, and thus N satisfies the Euler-Lagrange equation (4.7). Let us summarize.

Theorem B. (Euler-Lagrange equation for one integral constraint) Given an absolutely continuous $y = y(x)$ with bounded derivative that minimizes (or maximizes) the cost

$$J = \int_a^b L(x, y(x), y'(x)) \, dx \tag{5.15a}$$

5.5 Example Applications

subject to the constraint

$$C = \int_a^b M(x, y(x), y'(x)) \, dx = c_0, \tag{5.15b}$$

it is necessary that either the constraint integrand M satisfies the Euler-Lagrange equation (4.7), or, for some constant λ,

$$N = L + \lambda M \tag{5.16}$$

satisfies the Euler-Lagrange equation, that is,

$$\frac{\partial N}{\partial y} - \frac{d}{dx}\frac{\partial N}{\partial y'} = 0. \tag{5.17}$$

5.5 Example Applications

Example 1. (Dido's problem CVP 10). We are to maximize the area

$$J = \int_0^1 y \, dx \tag{5.18a}$$

enclosed by the curve $y = y(x)$ of length

$$\pi/2 = \int_0^1 \sqrt{1 + y'^2} \, dx, \tag{5.18b}$$

together with the line segment $[0, 1]$ of the x-axis. Note that it is impossible in this situation for $M = \sqrt{1 + y'^2}$ to satisfy the Euler-Lagrange equation (Exercise 5.2).

Applying the Euler-Lagrange equation (5.17) to

$$N = y + \lambda\sqrt{1 + y'^2} \tag{5.19}$$

yields that (Exercise 5.3)

$$\frac{y'}{\sqrt{1 + y'^2}} = \frac{x - h}{\lambda}, \tag{5.20}$$

with solutions (Exercise 5.4)

$$(x - h)^2 + (y - k)^2 = \lambda^2, \tag{5.21}$$

that is, the required curve $y = y(x)$ is an arc of a circle of radius λ with chord $[0, 1]$ (see Figure 5.2).

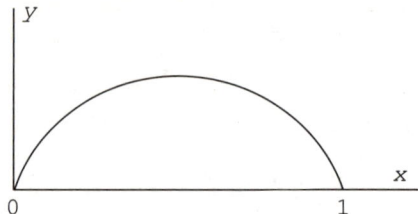

Figure 5.2 The Euler-Lagrange equation implies that $y = y(x)$ is a circular arc.

But only one of these circular arcs has length $\pi/2$, namely the semicircle. See Exercise 5.14.

Example 2. (CVP 12) What planar curve of arclength π encloses the largest area?

(The many technical details of the following example are left as a significant student project as Exercise 5.7.) We of course intuit that a circle of radius $1/2$ is the correct solution. In polar coordinates, we are to maximize the area enclosed by $r = r(\theta)$, that is, to maximize

$$J = \int_0^{2\pi} \frac{r^2 d\theta}{2} \tag{5.22a}$$

subject to the constraint

$$\pi = \int_0^{2\pi} \sqrt{r^2 + r'^2}\, d\theta. \tag{5.22b}$$

The arclength integrand cannot satisfy the Euler-Lagrange equation (Exercise 5.6). Thus we must instead have that the augmented Lagrangian satisfies the Euler-Lagrange equation:

$$\frac{d}{d\theta} \frac{\lambda r'}{\sqrt{r^2 + r'^2}} = r + \frac{\lambda r}{\sqrt{r^2 + r'^2}}, \tag{5.23}$$

where $\lambda < 0$. Certainly $r = -\lambda$ is a solution.

Introduce the intermediate parameter χ via the relation

$$\cos \chi = \frac{r'}{\sqrt{r^2 + r'^2}} = \frac{dr}{ds}, \tag{5.24a}$$

5.5 Example Applications

and hence
$$\sin \chi = \frac{r}{\sqrt{r^2 + r'^2}} = r\frac{d\theta}{ds}, \tag{5.24b}$$

where s is arclength. With this new parameter (5.23) becomes
$$\lambda \frac{d}{d\theta} \cos \chi = r + \lambda \sin \chi, \tag{5.25}$$

giving that
$$-r = \lambda[\sin \chi - \frac{d}{d\theta} \cos \chi] = \lambda[1 + \frac{d\chi}{d\theta}] \sin \chi. \tag{5.26}$$

By introducing the important parameter
$$\phi = \theta + \chi, \tag{5.27}$$

(5.26) becomes
$$-r = \lambda r \frac{d\theta}{ds} \frac{d\phi}{d\theta}, \tag{5.28}$$

and hence
$$\frac{d\phi}{ds} = -\frac{1}{\lambda}, \tag{5.29}$$

that is,

our sought for curve has constant curvature.

In more detail, note that
$$\cos \phi = \cos(\theta + \chi) = \cos \theta \cos \chi - \sin \theta \sin \chi$$
$$= \frac{d}{ds}(r \cos \theta) = \frac{dx}{ds}, \tag{5.30a}$$

and likewise
$$\sin \phi = \frac{d}{ds}(r \sin \theta) = \frac{dy}{ds}. \tag{5.30b}$$

Thus
$$\tan \phi = \frac{dy/ds}{dx/ds} = \frac{dy}{dx},$$

and so we recognize ϕ as the angle with the x-axis made by the tangent to our sought-for curve. Moreover, as is always the case when we parameterize with arclength,
$$\left(\frac{dx}{ds}\right)^2 + \left(\frac{dy}{ds}\right)^2 = \cos^2 \phi + \sin^2 \phi = 1.$$

See Figure 5.3.

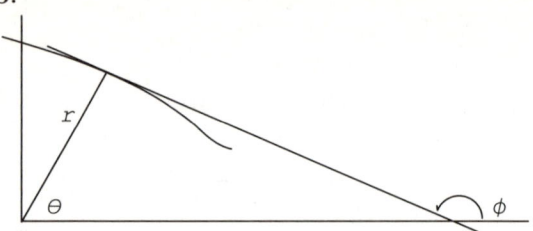

Figure 5.3 The angle of the tangent with the x-axis is ϕ.

By differentiating (5.30) we obtain

$$\frac{\sin \phi}{\lambda} = \frac{d^2x}{ds^2} \quad \text{and} \quad -\frac{\cos \phi}{\lambda} = \frac{d^2y}{ds^2}, \tag{5.31}$$

so that

$$\left(\frac{d^2x}{ds^2}\right)^2 + \left(\frac{d^2y}{ds^2}\right)^2 = \frac{1}{\lambda^2} = \kappa^2. \tag{5.32}$$

It remains to show that a plane curve with constant (nonzero) curvature κ is a circle.

Because the tangent 'velocity' vector $v = (dx/ds, dy/ds)$ is of constant length, the 'acceleration' vector $a = (d^2x/ds^2, d^2y/ds^2)$ is perpendicular to v and points inward. Thus the *center of curvature* is at

$$(h, k) = (x, y) + \frac{1}{\kappa} \frac{a}{|a|} = (x + \lambda^2(\frac{d^2x}{ds^2}), y + \lambda^2(\frac{d^2y}{ds^2}))$$

$$= (x + \lambda \sin \phi, y - \lambda \cos \phi),$$

which when differentiated becomes

$$\frac{d}{ds}(h, k) = (\frac{dx}{ds} - \cos \phi, \frac{dy}{ds} - \sin \phi) = (0, 0),$$

that is, the center (h, k) of curvature is fixed, and in fact

$$x = h - \lambda \sin \phi \quad \text{and} \quad y = k + \lambda \cos \phi.$$

Thus our curve enclosing maximal area is a circle of radius $r = -\lambda$ centered at (h, k).

5.5 Example Applications

Example 3. **(CVP 13)** Which surface of revolution of a given surface area encloses the maximum volume?

Let us do the simplest case: Revolve the curve $y = y(x)$ about the x-axis for $a \leq x \leq b$, where $y(a) = 0 = y(b)$. We are to maximize the volume

$$J = \pi \int_a^b y^2 \, dx \tag{5.33a}$$

while holding fixed the surface area

$$2\pi \int_a^b y\sqrt{1+y'^2} \, dx = \gamma. \tag{5.33b}$$

When a and b are free to move, the answer $y = y(x)$ is of course a semicircle—the maximum volume enclosed by a given surface area is enclosed by a sphere. But what if a and b are not free to move?

Applying the Euler-Lagrange equations (5.18) to $N = y^2 + \lambda y\sqrt{1+y'^2}$, we obtain

$$\lambda \frac{d}{dx} \frac{yy'}{\sqrt{1+y'^2}} = 2y + \lambda\sqrt{1+y'^2}. \tag{5.34}$$

Instead of proceding with this relationship, let us instead apply the "no x" result (3.15)[1] to N to obtain (Exercise 5.15)

$$y^2 + \frac{\lambda y}{\sqrt{1+y'^2}} = c. \tag{5.35}$$

Since $y(a) = 0 = y(b)$, the constant of integration in (5.35) is $c = 0$. Thus

$$y + \frac{\lambda}{\sqrt{1+y'^2}} = y + \lambda \frac{dx}{ds} = 0. \tag{5.36}$$

As in the previous example, introduce the parameter ϕ via the relations

$$\frac{dx}{ds} = \cos\phi \quad \text{and} \quad \frac{dy}{ds} = \sin\phi,$$

giving that

$$y + \lambda \cos\phi = 0.$$

---

[1] The "no x" result holds whenever the Euler-Lagrange equations hold.

Differentiating yields

$$\frac{dy}{ds} - \lambda \sin\phi \frac{d\phi}{ds} = \sin\phi - \lambda \sin\phi \frac{d\phi}{ds} = 0,$$

that is,

$$\frac{d\phi}{ds} = \frac{1}{\lambda},$$

and so, as in Example 2, because the curvature $d\phi/ds$ is constant, our optimal curve $y = y(x)$ is a circular arc. [Alternatively, square (5.36), solve for y', and integrate.]

5.6 Multiple Degrees of Freedom

Theorem B and its proof generalize to problems with multiple degrees of freedom. Assume that the Lagrangian L and constraint M are everywhere continuously differentiable.

Theorem C. Given a piecewise continuously differentiable path $q = q(t)$ in $I\!R^n$ that minimizes (or maximizes) the cost

$$J = \int_a^b L(t, q(t), \dot q(t))\, dt \qquad (5.37a)$$

and satisfies the constraint

$$C = \int_a^b M(t, q(t), \dot q(t))\, dt = c_0, \qquad (5.37b)$$

it is necessary that either the constraint integrand M satisfies the Lagrange equations (4.30), or, for some constant λ, the *augmented Lagrangian*

$$N = L + \lambda M \qquad (5.38)$$

satisfies the Lagrange equations, that is,

$$\frac{\partial N}{\partial q_k} - \frac{d}{dt}\frac{\partial N}{\partial \dot q_k} = 0, \qquad k = 1, 2, \ldots, n. \qquad (5.39)$$

Proof. Proceed as in the proof of Theorem B with perturbations of the form $q + \epsilon\eta + \delta\zeta$ (Exercise 5.18).

5.7 Nonintegral Constraints

Example 4. Which path of a given length through a force field requires the least work?

For example, suppose the plane is subjected to the force field

$$F(x,y) = y\mathbf{i} - x\mathbf{j}. \tag{5.40}$$

Which path $C : x = x(t), y = y(t)$ of length γ from (x_0, y_0) to (x_1, y_1) is the easiest to traverse? That is, we must minimize the cost

$$W = -\int F \cdot dv = -\int_C y\, dx - x\, dy = \int_a^b (x\dot{y} - y\dot{x})\, dt \tag{5.41a}$$

subject to

$$\int_a^b \sqrt{\dot{x}^2 + \dot{y}^2}\, dt = \gamma. \tag{5.41b}$$

Set

$$N = x\dot{y} - y\dot{x} + \lambda\sqrt{\dot{x}^2 + \dot{y}^2} \tag{5.42}$$

and impose the Lagrange equations to obtain (Exercise 5.33)

$$2\dot{y} = \lambda \frac{d}{dt} \frac{\dot{x}}{\sqrt{\dot{x}^2 + \dot{y}^2}}, \tag{5.43a}$$

$$-2\dot{x} = \lambda \frac{d}{dt} \frac{\dot{y}}{\sqrt{\dot{x}^2 + \dot{y}^2}}, \tag{5.43b}$$

which when integrated give

$$y - k = \frac{\lambda}{2} \frac{\dot{x}}{\sqrt{\dot{x}^2 + \dot{y}^2}}$$

and

$$-x + h = \frac{\lambda}{2} \frac{\dot{y}}{\sqrt{\dot{x}^2 + \dot{y}^2}}.$$

But this means (Exercise 5.33) that the path is the circular arc:

$$(x - h)^2 + (y - k)^2 = \left(\frac{\lambda}{2}\right)^2. \tag{5.44}$$

Choose h, k, R so that the arc has length γ and passes through the end points.

5.7 Non-Integral Constraints

Many important problems involve nonintegral constriants. Fortunately, we again have a result about the augmented Lagrangian similar to our Theorem C on integral constraints—but with one change.

Theorem D. Suppose the piecewise continuously differentiable curve $q = q(t)$ minimizes (maximizes) the cost

$$J = \int_a^b L(t, q, \dot{q})\, dt \qquad (5.45a)$$

subject to the constraint

$$M(t, q, \dot{q}) = 0 \qquad (5.45b)$$

on $a \leq t \leq b$, where both L and M are everywhere twice continuously differentiable. Moreover, assume that at each point $q(t_0)$ of the extremal path $q = q(t)$, at least one of the partial derivatives M_{q_j} is nonzero when evaluated at $q(t_0)$. Then locally, for all but a finite number of t, there exists a differentiable scalar function $\lambda = \lambda(t)$ so that the augmented Lagrangian

$$N = L + \lambda M \qquad (5.46)$$

satisfies the Lagrange equations

$$\frac{\partial N}{\partial q_k} - \frac{d}{dt}\frac{\partial N}{\partial \dot{q}_k} = 0, \qquad k = 1, 2, \ldots, n. \qquad (5.47)$$

Proof for the holonomic case. Assume the constraint is of *holonomic* form

$$M = M(q) = 0. \qquad (5.48)$$

Since $\nabla M \neq 0$, at each point q of the path $q = q(t)$, we may by compactness find finitely many open balls covering the path of $q = q(t)$, where by the implicit function theorem (Appendix A) we may solve $M(q) = 0$ for one variable in terms of the others, for example, $q_n = Q(q_1, \ldots, q_{n-1})$.

On the other hand, the curve $q = q(t)$ is also a local extremal, that is, the portion that $q = q(t)$ contributes toward the cost J on each $[t_0 - \delta, t_0 + \delta]$ must be optimal, otherwise more optimal solutions on

5.7 Nonintegral Constraints

subintervals could be pasted together to yield an even more favorable total cost J. So we may assume from the outset that $M_{q_n} \neq 0$ and we have a global solution of the constraint surface $M = 0$ given by

$$(q_1, \ldots, q_{n-1}, Q(q_1, \ldots, q_{n-1})) = (\tilde{q}, Q(\tilde{q})), \quad (5.49)$$

where Q is continuously differentiable on an open neighborhood of the path $\tilde{q}(t) = (q_1(t), \ldots, q_{n-1}(t))$ for $a \leq t \leq b$.

Let[2]

$$\Lambda = L(t, \tilde{q}, Q, \dot{\tilde{q}}, P), \quad (5.50a)$$

where

$$P = \sum_{j=1}^{n-1} \frac{\partial Q}{\partial q_j} \dot{q}_j. \quad (5.50b)$$

Since $q = q(t)$ is minimum (maximum) for the cost J subject to $M = 0$, so must $\tilde{q}(t)$ be minimum (maximum) for

$$J_0 = \int_a^b \Lambda(t, \tilde{q}, Q, \dot{\tilde{q}}, P)\, dt. \quad (5.51)$$

The Lagrange equations applied to the integrand of (5.51) yield that

$$\frac{\partial \Lambda}{\partial q_k} = \frac{d}{dt} \frac{\partial \Lambda}{\partial \dot{q}_k}, \qquad k = 1, \ldots, n-1,$$

that is,

$$\frac{\partial L}{\partial q_k} + \frac{\partial L}{\partial q_n} \frac{\partial Q}{\partial q_k} + \frac{\partial L}{\partial \dot{q}_n} \frac{\partial P}{\partial q_k} = \frac{d}{dt} \left[\frac{\partial L}{\partial \dot{q}_k} + \frac{\partial L}{\partial \dot{q}_n} \frac{\partial Q}{\partial q_k} \right]. \quad (5.52)$$

But taking the derivative of the second term on the right,

$$\frac{d}{dt} \frac{\partial L}{\partial \dot{q}_n} \frac{\partial Q}{\partial q_k} = \frac{\partial Q}{\partial q_k} \frac{d}{dt} \frac{\partial L}{\partial \dot{q}_n} + \frac{\partial L}{\partial \dot{q}_n} \frac{d}{dt} \frac{\partial Q}{\partial q_k}$$

$$= \frac{\partial Q}{\partial q_k} \frac{d}{dt} \frac{\partial L}{\partial \dot{q}_n} + \frac{\partial L}{\partial \dot{q}_n} \frac{\partial P}{\partial q_k}.$$

Thus (5.52) is actually (Exercise 5.20) the relation

$$\frac{\partial L}{\partial q_k} - \frac{d}{dt} \frac{\partial L}{\partial \dot{q}_k} + \frac{\partial Q}{\partial q_k} \left[\frac{\partial L}{\partial q_n} - \frac{d}{dt} \frac{\partial L}{\partial \dot{q}_n} \right] = 0,$$

---
[2]What follows is a special case of the proof of Theorem E.

that is,
$$\frac{\partial L}{\partial q_k} - \frac{d}{dt}\frac{\partial L}{\partial \dot{q}_k} + \lambda \frac{\partial M}{\partial q_k} = 0, \tag{5.53}$$

with
$$\lambda = \frac{1}{M_{q_n}}[\frac{d}{dt}L_{\dot{q}_n} - L_{q_n}]. \tag{5.54}$$

The last Lagrange equation corresponding to (5.53) for $k = n$ is an obvious identity.

Example 5. (CVP 14) What are the *geodesics* on a sphere? That is, what is the curve $x = x(t)$, $y = y(t)$, $z = z(t)$ of shortest length connecting two given points A and B on the unit sphere $x^2 + y^2 + z^2 = 1$? We of course know the answer: the shortest path is along great circles. Let us derive this analytically.

We are to minimize
$$J = \int_a^b \sqrt{\dot{x}^2 + \dot{y}^2 + \dot{z}^2}\, dt \tag{5.55a}$$

subject to the holonomic condition
$$x^2 + y^2 + z^2 = 1. \tag{5.55b}$$

Hence locally, for some $\lambda = \lambda(t)$ the Lagrange equations (5.47) become (Exercise 5.15)
$$\frac{d}{dt}\frac{v}{|v|} = 2\lambda r, \tag{5.56}$$

where $v = (\dot{x}, \dot{y}, \dot{z})$, $r = (x, y, z)$, and $|r| = 1$. But now employ a famous trick (see Exercise 5.19): The cross product $p = (v/|v|) \times r$ has time derivative
$$\frac{dp}{dt} = \frac{d(v/|v|)}{dt} \times r + \frac{v}{|v|} \times \frac{dr}{dt} = 2\lambda r \times r + \frac{v}{|v|} \times v = 0 + 0 = 0.$$

But this means that the geodesic curve must lie in a plane perpendicular to the constant vector p containing r (i.e., in a plane through the origin, A, and B). The curve is portion of a great circle.

5.8 Hamilton's Principle with Constraints

Will Hamilton's principle extend to systems with constraints? Usually in practice the constraints are *holonomic*, that is, of the form

$$M_i(q) = 0, \qquad i = 1, 2, \ldots, s. \tag{5.57}$$

The n previously independent configuration variables $q = (q_1, \ldots, q_n)$ are now no longer independent—a new independent set of independent configuration variables $\theta_1, \ldots, \theta_r$, with $r < n$, must be found.

Example 6. (The spherical pendulum) Consider the motions of a mass at one end of a massless rod of length 1, where the other end of the rod fixed at the origin but free to pivot in all directions,[3] that is, the system with Lagrangian

$$L = m\frac{\dot{x}^2 + \dot{y}^2 + \dot{z}^2}{2} - mgz, \tag{5.58a}$$

subject to the constraint

$$M(x, y, z) = x^2 + y^2 + z^2 - 1 = 0. \tag{5.58b}$$

The strategy is to find *two* new independent configuration variables. For instance, use the modified spherical coordinates

$$x = \sin\phi \cos\theta, \ y = \sin\phi \sin\theta, \ z = -\cos\phi,$$

where ϕ is the angle made with the negative z-axis.

Hamilton's Principle. Using the constraints, find new independent configuration variables and rewrite the Lagrangian in terms of these new variables. Then each mechanical trajectory must satisfy the Lagrange equations with respect to these new configuration variables.

Often these steps can actually be carried out. More often the computational details are overwelming. Let us search for a shortcut.

---

[3]Or think of a small mass sliding around the inside of the unit sphere under the influence of gravity, assuming motions never reach above the equator (where the mass might fall off the surface). If large motions are allowed, this becomes the classic example of a nonholonomic constraint.

Chapter 5. Constrained Problems

Example 7. The inclined plane.

Consider a mass $m = 1$ located at $(x(t), y(t))$ that is free to move within the xy-plane, influenced only by the downward acceleration g of gravity. By rescaling we may assume $g = 1$. The Lagrangian thus becomes

$$L = \frac{\dot{x}^2 + \dot{y}^2}{2} - y. \tag{5.59}$$

But now impose the constraint that the mass must slide along the inclined plane $y = x$, that is,

$$M(x, y) = y - x = 0. \tag{5.60}$$

Solution 1. We eliminate the dependent configuration variable y using the constraint $M = 0$ to obtain the new Lagrangian

$$\Lambda = \frac{\dot{x}^2 + \dot{x}^2}{2} - x = \dot{x}^2 - x; \tag{5.61}$$

then apply the Lagrange equation to obtain

$$-1 = 2\ddot{x},$$

that is,

$$y = x = -t^2/4 + \dot{x}(0)t + x(0). \tag{5.62}$$

Solution 2. Insisting that $N = L + \lambda(t)M$ satisfies the Lagrange equations yields that

$$-\lambda = \frac{\partial N}{\partial x} = \frac{d}{dt}\frac{\partial N}{\partial \dot{x}} = \ddot{x}, \tag{5.63a}$$

$$-1 + \lambda = \frac{\partial N}{\partial y} = \frac{d}{dt}\frac{\partial N}{\partial \dot{y}} = \ddot{y}. \tag{5.63b}$$

Because $y = x$, we see $\lambda = 1/2$, which gives the correct motions

$$x = -\lambda t^2/2 + \dot{x}(0)t + x(0), \tag{5.64a}$$

$$y = -(1 - \lambda)t^2/2 + \dot{y}(0)t + y(0), \tag{5.64b}$$

when $\lambda(t) = 1/2$. So there appears to be a hidden Lagrange multiplier result for these constrained mechanical systems.

5.8 Hamilton's Principle with Constraints

Example 8. The planar pendulum revisited.

Let us continue the previous example with Lagrangian

$$L = \frac{\dot{x}^2 + \dot{y}^2}{2} - y \tag{5.65a}$$

but now with constraint

$$M = x^2 + y^2 - 1 = 0, \tag{5.65b}$$

a pendulum of arm length $a = 1$.

Solution 1. Use the angle θ made with the negative y-axis as the single independent degree of freedom (Figure 5.4):

$$x = \sin\theta, \qquad y = -\cos\theta. \tag{5.66}$$

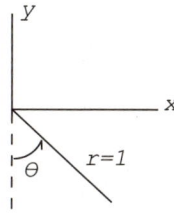

Figure 5.4

We are immediately led to the correct model (Exercise 5.22) by applying the Lagrange equation:

$$\ddot{\theta} = -\sin\theta. \tag{5.67}$$

Solution 2. Applying the Lagrange equations to $N = L + \lambda(t)M$ yields

$$2\lambda x = \frac{\partial N}{\partial x} = \frac{d}{dt}\frac{\partial N}{\partial \dot{x}} = \ddot{x}, \tag{5.68a}$$

$$-1 + 2\lambda y = \frac{\partial N}{\partial y} = \frac{d}{dt}\frac{\partial N}{\partial \dot{y}} = \ddot{y}. \tag{5.68b}$$

These equations are not much help; it is difficult to deduce the correct model without making the substitutions (5.66)—try it! Work Exercise 5.23.

84 Chapter 5. Constrained Problems

Question. Is there a Lagrange multiplier result for constrained mechanical systems?

Theorem E. Suppose an n-degree-of-freedom mechanical system with Lagrangian $L(t, q, \dot{q})$ is then constrained by $s < n$ independent and consistent holonomic constraints $M_i(q) = 0$, $i = 1, 2, \ldots, s$.

That is, assume there are points q common to all constraint surfaces $M_i(q) = 0$, and that at each such point q^0,

(a) all $\dfrac{\partial M_j}{\partial q_i}$ are continuous on an open neighborhood of q^0, and

(b) the rank at q^0 of the matrix $\left[\dfrac{\partial M_j}{\partial q_i}\right]$ is s.

Then for each mechanical trajectory $q = q(t)$ with $q^0 = q(t_0)$ common to all constraint surfaces, there are s parameters $\lambda_i(t)$, $i = 1, 2, \ldots, s$, differentiable near $t = t_0$ such that the augmented Lagrangian

$$N = L(t, q, \dot{q}) + \lambda_1(t) M_1(q) + \cdots + \lambda_s(t) M_s(q) \tag{5.69}$$

satisfies the Lagrange equations

$$\frac{\partial N}{\partial q_k} - \frac{d}{dt}\frac{\partial N}{\partial \dot{q}_k} = 0, \qquad k = 1, 2, \ldots, n. \tag{5.70}$$

Proof. Suppose for the moment that we have found $r = n - s$ new independent configuration variables $\theta_1, \ldots, \theta_r$ dictated by the s constraints $M_i(q) = 0$. We rewrite the Lagrangian in terms of these new variables as

$$\Lambda(t, \theta, \dot{\theta}) = L(t, q(\theta), p(\theta, \dot{\theta})), \tag{5.71}$$

where

$$p_k = \sum_{j=1}^{r} \frac{\partial q_k}{\partial \theta_j} \dot{\theta}_j. \tag{5.72}$$

By Hamilton's principle, any mechanical trajectory of the constrained system must satisfy the Lagrange equations with respect to the new configuration variables θ. But

$$\frac{\partial \Lambda}{\partial \theta_k} = \sum_{j=1}^{n} \frac{\partial L}{\partial q_j}\frac{\partial q_j}{\partial \theta_k} + \sum_{j=1}^{n} \frac{\partial L}{\partial \dot{q}_j}\frac{\partial p_j}{\partial \theta_k} \tag{5.73}$$

5.8 Hamilton's Principle with Constraints

and

$$\frac{d}{dt}\frac{\partial \Lambda}{\partial \dot\theta_k} = \sum_{j=1}^{n} \frac{d}{dt}\left(\frac{\partial L}{\partial \dot q_j}\frac{\partial p_j}{\partial \dot\theta_k}\right)$$

$$= \sum_{j=1}^{n} \frac{\partial q_j}{\partial \theta_k}\frac{d}{dt}\frac{\partial L}{\partial \dot q_j} + \sum_{j=1}^{n} \frac{\partial L}{\partial \dot q_j}\frac{d}{dt}\frac{\partial q_j}{\partial \theta_k}$$

$$= \sum_{j=1}^{n} \frac{\partial q_j}{\partial \theta_k}\frac{d}{dt}\frac{\partial L}{\partial \dot q_j} + \sum_{j=1}^{n} \frac{\partial L}{\partial \dot q_j}\frac{\partial p_j}{\partial \theta_k}. \qquad (5.74)$$

Equating (5.73) and (5.74) yields the relation

$$\sum_{j=1}^{n} \frac{\partial q_j}{\partial \theta_k}\frac{\partial L}{\partial q_j} = \sum_{j=1}^{n} \frac{\partial q_j}{\partial \theta_k}\frac{d}{dt}\frac{\partial L}{\partial \dot q_j},$$

that is, the column $(L_{q_1} - \frac{d}{dt}L_{\dot q_1}, \ldots, L_{q_n} - \frac{d}{dt}L_{\dot q_n})^T$ is in the null space of

$$Q = \begin{bmatrix} \frac{\partial q_1}{\partial \theta_1} & \frac{\partial q_2}{\partial \theta_1} & \cdots & \frac{\partial q_n}{\partial \theta_1} \\ \cdot & \cdot & \cdots & \cdot \\ \frac{\partial q_1}{\partial \theta_r} & \frac{\partial q_2}{\partial \theta_r} & \cdots & \frac{\partial q_n}{\partial \theta_r} \end{bmatrix}. \qquad (5.75)$$

On the other hand, by the chain rule, since $M_i(q(\theta)) \equiv 0$,

$$0 = \frac{\partial M_i}{\partial \theta_k} = \sum_{j=1}^{n} \frac{\partial q_j}{\partial \theta_k}\frac{\partial M_i}{\partial q_j}, \qquad (5.76)$$

and hence the null space of Q contains the s (independent) columns $\alpha_j = (\partial M_j/\partial q_1, \ldots, \partial M_j/\partial q_n)^T$. Assume Q has full rank r. Then the columns α_j must form a null-space basis since $r + s = n$. As a consequence, the Lagrange equations

$$-L_{q_k} + \frac{d}{dt}L_{\dot q_k} = \lambda_1 \partial M_1/\partial q_k + \cdots + \lambda_s \partial M_s/\partial q_k \qquad (5.77)$$

obtain. By Cramer's rule, the λ_i are locally differentiable (Exercise 5.27). Thus the augmented Lagrangian N locally satisfies the Lagrange equations. It is time to construct the variables θ where we will see that Q has rank r (Exercise 5.27).

Because the $n \times s$ matrix

$$M = \left[\frac{\partial M_j}{\partial q_i}\right] \qquad (5.78)$$

has full rank s, we may renumber the q_i so that the upper left $s \times s$ minor has rank s (Exercise 5.26). But then the Jacobian of the map

$$(q_1, q_2, \ldots, q_n) \longmapsto (M_1, M_2, \ldots, M_s, q_{s+1}, q_{s+2}, \ldots, q_n) \qquad (5.79)$$

is nonzero, and hence (by Theorem A) the mapping has a local differentiable inverse. This means we may take $\theta_i = q_{s+i}$ as the independent local configuration variables to locally satisfy $M_i(q(\theta)) = 0$. (These are almost never the best physical choices.) All necessary remaining details are obviated (Exercise 5.27).

The same proof and result hold not only for mechanical systems, but for any constrained Lagrangian.

Exercises

5.1 Show that if $C_\epsilon(0,0) = 0$ in (5.11) for all η, then M satisfies the Euler-Lagrange equation (3.7).

5.2 Show that in Dido's problem (Example 1), the constraint integrand satisfies the Euler-Lagrange equation only when the *minimum* area (zero) is attained (see CVP 1).

5.3 Verify (5.20).

5.4 Verify (5.21).

5.5 Verify (5.22b), that in polar coordinates the differential of arclength becomes
$$ds = \sqrt{r^2 + r'^2}\, d\theta.$$

5.6 Show that if the constraint integrand of (5.22b) in Example 2 satisfies the Euler-Lagrange equation, then the curve is a straight line.

5.7 **(Project)** Verify *all* computational and technical details of Example 2.

Outline: First verify that all calculations are formally correct. Then settle the many inherent technical issues (e.g., reparametrize by arclength, vertical tangents, the existence of the derivatives taken, the applications of the chain rule and implicit function theorem).

Exercises

5.8 Find a continuously differentiable curve $y = y(x)$ with $y(0) = 0 = y(1)$ that minimizes
$$J = \int_0^1 y'^2 \, dx$$
subject to the constraint
$$\int_0^1 y^2 \, dx = 1.$$
Answer: $y = \sqrt{2} \sin \pi x$.

5.9 Show that there is no twice-differentiable curve $y = y(x)$ with $y(0) = 0 = y(1)$ that minimizes the integral
$$J = \int_0^1 y'^2 \, dx$$
yet satisfies
$$\int_0^1 y'^2 y^2 \, dx = 1.$$
Hint: Apply the "no x" result to M and then to the augmented Lagrangian N.

5.10 Find a piecewise continuously differentiable curve $y = y(x)$ with $y(0) = 0 = y(1)$ that minimizes the integral
$$J = \int_0^1 y'^2 \, dx$$
subject to
$$\int_0^1 y'^2 y^2 \, dx = 1.$$

5.11 In the solution to Dido's problem in Example 1, the semicircular solution has a vertical tangent at $x = 0$ and at $x = 1$, thus violating several hypotheses. Repair this solution.

5.12 Deduce the **implicit function theorem** from Theorem A: Suppose
$$F(x, t) = (f_1(x, t), f_2(x, t), \ldots, f_n(x, t))$$
is a vector-valued function of the vector variables $x = (x_1, x_2, \ldots, x_n)$ and $t = (t_1, t_2, \ldots, t_r)$. Suppose that F is

continuously differentiable on a neighborhood of (x^0, t^0) and that the *Jacobian* with respect to x is nonzero at $x = (x^0, t^0)$, that is,
$$\det\left[\frac{\partial f_i}{\partial x_j}\right]|_{(x^0,t^0)} \neq 0.$$
Then there exists an open neighborhood U of t^0 and continuously differentiable functions $x = x(t)$ on U such that
$$F(x(t), t) = F(x^0, t^0)$$
for all t in U.

Hint: Set $f_{n+j}(x,t) = t_j$ and apply the inverse function theorem (Theorem A).

5.13 Deduce without computation the result of Example 2 from the result of Example 1.

5.14 Verify (5.34) and (5.35).

5.15 Verify (5.56).

5.16 Model the motions of a mass sliding down $y = x^2$.

5.17 Model the motions of a tiny spherical mass *rolling* down $y = x^2$.

5.18 Prove Theorem C by aping the proof of Theorem B.

5.19 Prove that the orbit of a particle subject solely to acceleration directed toward some single central point O must lie in a fixed plane. See Kepler's first law.

Hint: Differentiate 'angular momentum' $p = v \times r$.

5.20 Verify (5.53)–(5.54). Check that (5.53) is a trivial identity for $k = n$.

Hint: Take the partial derivative of (5.48).

5.21 Write the Lagrange equations for the augmented Lagrangian of the spherical pendulum (Example 6, §5.5).

5.22 Verify (5.67).

5.23 Attempt to deduce the model for the planar pendulum (CVP 8) from the Lagrange multiplier approach (5.70).

Exercises

5.24 Model the motions of a elliptical pendulum (i.e., a mass constrained to lie on the ellipse $x^2/a^2 + y^2/b^2 = 1$ under the affect of downward gravity). It is enough to do the nondimensional version: $m = g = 1$.

Hint: Let $x = a \sin \theta$ and $y = -b \cos \theta$.

5.25 In the previous exercise, suppose the mass is a ball bearing *rolling* within the ellipse without slipping. What is the model now?

5.26 Establish that by resubscripting the q in the proof of Theorem E, we may assume that the upper left minor

$$\frac{\partial M_j}{\partial q_i}, \quad 1 \leq i, j \leq s,$$

is nonsingular. Note the similarity to Exercise 1.17.

5.27 Fill in all the details at the end of the proof of Theorem E. In particular, determine under what conditions are the λ_k are (locally) differentiable.

5.28 Generalize and prove Theorem D using the ideas of the proof of Theorem E.

5.29* Generalize and prove Theorems D and E for nonholonomic constraints $M_i(t, q, \dot{q}) = 0, \ i = 1, 2, \ldots, s$.

5.30 Find the path of least work done in directly crossing a stream against a constant current.

5.31 Find the path taken by a drake as he swims toward a duck on the opposite bank against constant current. As is observed, his heading is always toward her. How much longer will his crossing take in comparison to a more intelligent strategy?

5.32 Turn Dido's problem around—find the closed curve of least perimeter that encloses an area γ.

Hint: Use polar coordinates to obtain a cousin of (5.23).

5.33 Verify (5.43) and (5.44).

5.34 Track through the method of Example 4 for a conservative field, for instance, for $F = y\mathbf{i} + x\mathbf{j}$. How does the mathematics reflect that all curves of the correct length will work?

5.35 Suppose the curve $y = y(x)$ minimizes the cost

$$J = \int_a^b L(x, y, y') \, dx,$$

where the lower limit a is fixed but where the upper limit b must satisfy $y(b) = g(b)$ for some prescribed curve $y = g(x)$. Prove that not only does the Euler-Lagrange equation (4.7) hold, but in addition we have the *transversality condition*

$$L(b, y(b), y'(b)) + (g'(b) - y'(b)) \frac{\partial L}{\partial y'}(b, y(b), y'(b)) = 0.$$

Outline: In the usual way, show that L satisfies the Euler-Lagrange equation. Next, consider the perturbed curve $y = y(x) + \eta(x, \epsilon)$, where

$$\eta(x, \epsilon) = (x - a)(g(b + \epsilon) - y(b + \epsilon))/(b + \epsilon - a),$$

yielding the perturbed cost

$$J(\epsilon) = \int_a^{b+\epsilon} L(x, y(x) + \eta(x, \epsilon), y'(x) + \eta_x(x, \epsilon)) \, dx$$

with $J(\epsilon) \geq J(0) = J$. Apply the differentiation formula

$$\frac{d}{dy} \int_a^y f(x, y) \, dx = f(y, y) + \int_a^y \frac{\partial}{\partial y} f(x, y) \, dx.$$

5.36 Formulate and solve the problem of geodesic curves on the surface $M(x, y, z) = 0$. See Example 5.

Answer: $(\dot{x}/\dot{s})^\cdot / M_x = (\dot{y}/\dot{s})^\cdot / M_y = (\dot{z}/\dot{s})^\cdot / M_z$.

5.37 Prove that *a strict inequality constraint is no constraint at all*. That is, if the constraint $C = \gamma$ in (5.1b) is replaced by $C > \gamma$, then a local minimum of the new constrained problem is a local minimum of the unconstrained problem (5.1a).

5.38 Carefully solve for the shape of a hanging string of length γ; solve the catenary problem (3.19).

Chapter 6

Extremal Surfaces

Rather than searching for curves that minimize cost functionals, in this chapter we search for *surfaces* that will minimize a given cost functional. This leads naturally to minimal surfaces, Plateau's problem, stable fluid flows, Schrödinger's equation, eigenvalue problems, and the Rayleigh-Ritz numerical method.

6.1 A Soap Film (CVP 15)

What surface $z = z(x,y)$ with planar boundary curve $x^2 + y^2 = 1$, $z = 0$, has the least surface area? Put more vividly, since a soap film minimizes its potential energy, what shape will a soap film take with this circular frame?

The answer is of course the planar surface $z = 0$. Let us derive this analytically.

The surface area is

$$A = \int\int_{x^2+y^2\leq 1} \sqrt{1 + z_x^2 + z_y^2}\; dx\, dy. \qquad (6.1)$$

Going over to more general notation, this cost is in the form

$$J = \int_\Omega L(x, y, z, z_x, z_y)\, d\Omega. \qquad (6.2)$$

We proceed in the now familiar way by perturbing our extremal surface by an arbitrary continuously differentiable function $\eta = \eta(x,y)$ that vanishes on the boundary $\partial\Omega$ of Ω. Set

$$J(\epsilon) = \int_\Omega L(x, y, z + \epsilon\eta, z_x + \epsilon\eta_x, z_y + \epsilon\eta_y)\, d\Omega. \qquad (6.3)$$

Since $z = z(x,y)$ yields minimal cost $J(0)$,

$$0 = J'(0) = \int_\Omega [L_z\eta + L_{z_x}\eta_x + L_{z_y}\eta_y]\, d\Omega$$

$$= \int_\Omega [L_z - \frac{\partial}{\partial x} L_{z_x} - \frac{\partial}{\partial y} L_{z_y}] \eta \, d\Omega + \int_\Omega [\frac{\partial}{\partial x}(L_{z_x}\eta) + \frac{\partial}{\partial y}(L_{z_y}\eta)] \, d\Omega. \quad (6.4)$$

But the *divergence theorem* guarantees that for any continuously differentiable vector field F on Ω, the flux of F leaving through the boundary $\partial \Omega$ is the integral of the divergence of F over Ω (Appendix C):

$$\int_{\partial\Omega} F \cdot \mathbf{n} \, d\partial\Omega = \int_\Omega \nabla \cdot F \, d\Omega. \quad (6.5)$$

Thus by specializing to $F = (L_{z_x}\eta, L_{z_y}\eta)$, the second integral of (6.4) must vanish since η vanishes on the boundary $\partial\Omega$. But then, because η is arbitrary, by Exercise 6.1,

$$L_z - \frac{\partial}{\partial x} L_{z_x} - \frac{\partial}{\partial y} L_{z_y} = 0. \quad (6.6)$$

With little change to the preceding argument, we have proved an analog to the Euler-Lagrange equation but now for surfaces (Exercise 6.2).

Theorem A. (Lagrange, 1760) Suppose a surface is given by a vector $x = x(u,v)$ that is a continuously differentiable function of its parameters u, v ranging over a domain (open, connected subset) Ω of the plane. Suppose $x = x(u,v)$ minimizes (maximizes) the cost functional

$$J = \int_\Omega L(u, v, x, x_u, x_v) \, d\Omega, \quad (6.7)$$

where L is continuously differentiable. Then for $k = 1, 2, 3, \ldots, n$,

$$L_{x_k} - \frac{\partial}{\partial u} L_{x_{ku}} - \frac{\partial}{\partial v} L_{x_{kv}} = 0, \quad (6.8)$$

(where x_{ku} denotes the partial derivative with respect to u of the k-th coordinate function x_k of x, etc). The result holds regardless of whether or not the surface is specified along $\partial\Omega$.

Example 1. Let us complete CVP 15 begun previously—to find the surface $z = z(x,y)$ of least area bounded by the unit circle lying in the xy-plane. By (6.8),

$$\frac{\partial}{\partial z}\sqrt{1 + z_x^2 + z_y^2}$$

6.1 A Soap Bubble (CVP 14)

$$= \frac{\partial}{\partial x}\frac{\partial}{\partial z_x}\sqrt{1+z_x^2+z_y^2} + \frac{\partial}{\partial y}\frac{\partial}{\partial z_y}\sqrt{1+z_x^2+z_y^2}, \qquad (6.9)$$

that is,

$$0 = \frac{\partial}{\partial x}\frac{z_x}{\sqrt{1+z_x^2+z_y^2}} + \frac{\partial}{\partial y}\frac{z_y}{\sqrt{1+z_x^2+z_y^2}}. \qquad (6.10)$$

Expanding out (6.10) yields (Exercise 6.8) the famous

$$2H = \frac{z_{xx}(1+z_y^2) - 2z_x z_{xy} z_y + z_{yy}(1+z_x^2)}{(1+z_x^2+z_y^2)^{3/2}} = 0, \qquad (6.11)$$

where H is the *mean curvature* of the surface $z = z(x,y)$, which in this case, for our surface of minimum area, is zero.

Let us employ an ingenious trick from the proof of the maximal principle for the heat equation. Set $w = z + \epsilon x^2$ with $\epsilon > 0$. Then the mean curvature equation (6.11) becomes

$$(w_{xx} - 2\epsilon)(1+w_y^2) - 2(w_x - 2\epsilon x)w_{xy}w_y + w_{yy}(1+(w_x - 2\epsilon x)^2) = 0. \quad (6.12)$$

At an interior maximum of w, we must have $w_x = w_y = 0$ and $w_{xx}, w_{yy} \leq 0$. Thus at such an interior point, (6.12) becomes

$$(w_{xx} - 2\epsilon) + w_{yy}(1 + 4\epsilon^2 x^2) = 0, \qquad (6.13)$$

an impossibility since the first term is negative, the second nonpositive. Thus the maximum of w must occur on the boundary where $z \equiv 0$, giving that $w \leq \epsilon$ on the boundary, hence everywhere. Thus $z + \epsilon x^2 = w \leq \epsilon$, and so $z \leq \epsilon(1-x^2) \leq \epsilon$. Hence $z \leq 0$ since ϵ is arbitrary. But (6.11) must hold when z is replaced by $-z$. Thus our surface of minimum area is flat.

Remark. We have proved much more. We may translate any surface $z = z(x,y)$ of minimum area bounded by a horizontal planar curve C until C lies in the xy-plane, then contract until C lies within the unit disk, and thereby conclude by the preceding argument that the surface is planar. We may rotate or reparametrize other surfaces that have planar boundary curves C until they are in the *Monge* form $(u, v, f(u, v))$ with C horizontal. In fact, cut any surface of minimum area with a plane—any resulting bubble is flat. These surfaces of minimum area are very rare because they are *minimal surfaces*, a surface where at each point the mean curvature is zero.

What is the mean curvature of a surface S at a point p? Let N be the unit normal to the surface S at p. Choose any two perpendicular planes containing the normal N and intersect both with the surface S to obtain two perpendicular curves through p that lie in the surface S. The *normal curvature* $\kappa = N \cdot dT/ds$ of each curve is (within a sign) the reciprocal of the radius of the best fitting circle at p, where T is the curve's unit tangent (i.e., its 'velocity' when parametrized with respect to arclength; see CVP 12 in §5.5 and Figure 6.1). The mean curvature is the average of the two normal curvatures of the two perpendicular curves at their point of intersection p, that is,

$$H = \frac{\kappa_1 + \kappa_2}{2}. \tag{6.14}$$

So a minimal surface ($H = 0$) must at each point bend in two opposite directions—it must look like a saddle at each point.

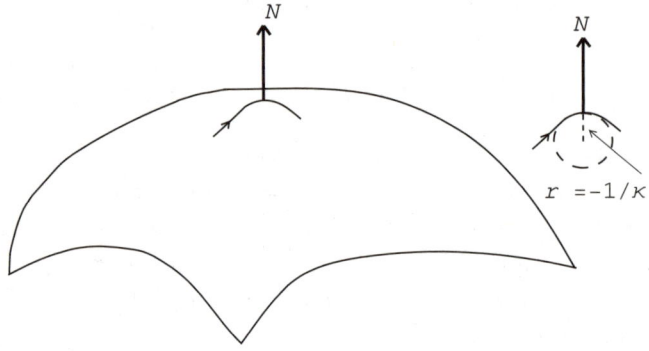

Figure 6.1 The normal curvature κ is within a sign the reciprocal of the radius r of curvature (i.e., the radius of the best fitting circle in the plane through the normal N cutting out the curve).

Plateau's Problem. Given an arbitrary boundary curve C, what shape will be assumed by a soap bubble bounded by C?

This old question was resolved in 1930 by J. Douglas and one of our teachers, Tibor Radó [Chen]. Experimentation is intuition building (Exercise 6.4).

6.1 A Soap Bubble (CVP 14)

Alert. If a soap bubble encloses a volume, it will contract until surface tension is balanced by interior pressure. The resulting surface (a sphere) will not have mean curvature 0. There are surfaces of mean curvature 0 that are not of minimum area. For the physics and mathematics underlying soap film and bubble problems, consult the wonderful book of John Oprea, *The Mathematics of Soap Films*, or [Osserman].

Example 2. **(CVP 16)** What is the surface of minimal area between the two boundary curves $x^2 + y^2 = 1$, one at $z = 0$, the other at $z = 1$? See Figure 6.2.

Several moments of reflection will reveal that the minimal surface is a surface of revolution about the z-axis (Exercise 6.6). Its area is

$$A = \int_0^1 2\pi x \sqrt{1 + (dx/dz)^2} \, dz. \qquad (6.15)$$

Look familiar? This is CVP 3 solved in §4.2, the catenary. Thus the surface is given by rotating

$$\cosh\left(\frac{z - 1/2}{c}\right) = \frac{x}{c} \qquad (6.16a)$$

about the z-axis, where

$$\cosh\left(\frac{1}{2c}\right) = \frac{1}{c}. \qquad (6.16b)$$

The graph of this catenoid is shown in Figure 6.2, obtained via the *Mathematica* script 6.1:

Routine 6.1 catenoid

```
FindRoot[Cosh[1/(2*c)] == 1/c, {c, 1}];
c = c/.%
x[z_, th_] := c*Cosh[(z - 0.5)/c]*Cos[th];
y[z_, th_] := c*Cosh[(z - 0.5)/c]*Sin[th];
ParametricPlot3D[{x[z, th], y[z, th], z}, {z, 0, 1}, {th, 0, 2*Pi}]
```

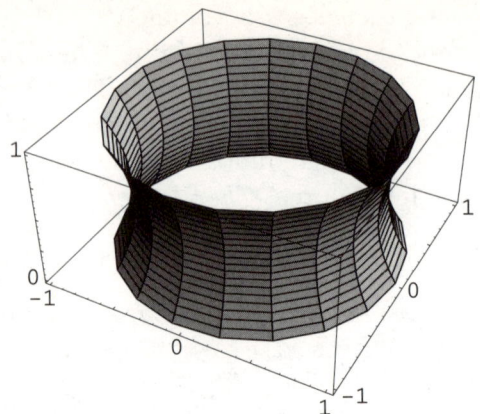

Figure 6.2 One of the rare minimal surfaces, a *catenoid*, obtained by revolving a catenary about the *z*-axis.

6.2 Stable Flows

Example 3. (**CVP 17**) Suppose V is the (vector) velocity of a steady irrotational fluid flow within a planar region Ω. Assume the density ρ is constant so that the net flow into the region must be 0. Because the flow is irrotational, by Stokes's theorem [Folland], the velocity V has a scalar potential v, that is,

$$V = -\nabla v, \qquad (6.17)$$

and so the total kinetic energy T of the fluid within the region is determined by

$$2T = \int_\Omega V \cdot V \, dm = \int_\Omega V \cdot V \, \rho \, d\Omega = \rho \int_\Omega v_x^2 + v_y^2 \, d\Omega, \qquad (6.18)$$

which by the Lagrange theorem (6.8) is stationary exactly when (Exercise 6.31)

$$v_{xx} + v_{yy} = 0, \qquad (6.19)$$

that is, when the potential v is *harmonic*. This also follows via the divergence theorem (Exercise 6.12). The fluid selects a stable flow.

In general,

Result A. $\int_\Omega (\nabla u)^2 \, d\Omega$ is stationary only when $\nabla^2 u = 0$ on Ω.

In many distributed mechanical problems, potential energy appears (within units) as a *Dirichlet norm*

$$[u, u] = \int_\Omega \nabla u \cdot \nabla u \, d\Omega, \qquad (6.20)$$

where u is displacement from equilibrium (Exercise 6.14). A stable potential energy configuration will by Result A yield a harmonic u.

6.3 Schrödinger's Equation (CVP 18)

The following is in the spirit of Schrödinger's first derivation [Weinstock]. Suppose a single electron of mass m with total energy E is orbiting its nucleus situated at the origin. At each position (x, y, z) in its orbit the electron has an $1/r$ electrostatic potential energy $V(x, y, z)$ and kinetic energy $T(x, y, z) = p^2/2m = E - V$. Let $S(x, y, z)$ be a 'potential' for p^2 [i.e., $(\nabla S)^2 = p^2$ (Exercise 6.23)]. Then

$$\frac{1}{2m}(S_x^2 + S_y^2 + S_z^2) + V - E = 0. \qquad (6.21)$$

Make the substitution

$$S = \hbar \log \psi \qquad (6.22)$$

to obtain

$$\frac{\hbar^2}{2m}(\psi_x^2 + \psi_y^2 + \psi_z^2) + (V - E)\psi^2 = 0. \qquad (6.23)$$

Because the orbit is stable, any solution to (6.23) must make stationary the integral

$$J = \int_{R^3} \frac{\hbar^2}{2m}(\psi_x^2 + \psi_y^2 + \psi_z^2) + (V - E)\psi^2 \, dx \, dy \, dz, \qquad (6.24)$$

which by the 3-variable version of the Lagrange equation (Exercise 6.13) yields (Exercise 6.32)

$$-\frac{\hbar^2}{2m}\nabla^2\psi + V\psi = E\psi, \qquad (6.25)$$

the famous eigenvalue equation for the stationary states of the wave function ψ of the hydrogen atom. See [Weinstock], [van der Waerden], and [MacCluer, 2004] for more detail.

6.4 Eigenvalue Problems

Eigenvalue problems can be restated variationally, then attacked numerically via the powerful Rayleigh-Ritz and Galerkin numerical methods. Let us begin abstractly.

A matrix, integral, or differential linear operator A has the scalar λ as an *eigenvalue* if for some nonzero vector f,

$$Af = \lambda f. \tag{6.26}$$

Note that the *eigenvector (eigenfunction)* f is determined only within scalar multiples and so we are at liberty to insist that f be of *norm* 1, that is,

$$\langle f, f \rangle = 1,$$

where the *inner product*

$$\langle g, h \rangle = \int_\Omega g(x) h(x) \, d\Omega \tag{6.27}$$

when g, h are (realvalued) functions on some domain $\Omega \subset \mathbb{R}^n$, or

$$\langle g, h \rangle = h^T g, \tag{6.28}$$

when g, h are (real) column vectors. An operator A is *self-adjoint* (*Hermitian*) if

$$\langle Ag, h \rangle = \langle g, Ah \rangle \tag{6.29}$$

for all g, h in the domain of A. See Exercises 6.15–6.18. We may take the inner product with f on both sides of (6.26) to obtain

$$\langle Af, f \rangle = \langle \lambda f, f \rangle = \lambda \langle f, f \rangle.$$

Result B. Suppose A is self-adjoint. If f is an eigenfunction belonging to the eigenvalue λ, then f makes stationary the cost

$$J = \langle Ag, g \rangle - \lambda \langle g, g \rangle. \tag{6.30}$$

Proof. Suppose $Af = \lambda f$. When we slightly perturb the eigenfunction f in the direction η, for η an arbitrary element of the domain of A, we see that the cost

$$J(\epsilon) = \langle A(f + \epsilon\eta), f + \epsilon\eta \rangle - \lambda \langle f + \epsilon\eta, f + \epsilon\eta \rangle$$

$$= \langle Af, f\rangle + \epsilon\langle Af, \eta\rangle + \epsilon\langle A\eta, f\rangle + \epsilon^2\langle A\eta, \eta\rangle - \lambda\langle f, f\rangle - 2\epsilon\lambda\langle f, \eta\rangle - \epsilon^2\lambda\langle \eta, \eta\rangle$$
$$= \langle (A-\lambda)f, f\rangle + 2\epsilon\langle (A-\lambda)f, \eta\rangle + \epsilon^2\langle (A-\lambda)\eta, \eta\rangle$$
$$= \epsilon^2\langle (A-\lambda)\eta, \eta\rangle. \tag{6.31}$$

Conversely, suppose $f \neq 0$ is critical for the cost (6.30). Then it follows (Exercise 6.24) that λ is an eigenvalue of A belonging to the eigenfunction f, provided Af is in the domain of A.

6.5 Rayleigh-Ritz Numerics

Our goal is to approximate the eigenvalues λ and eigenfunctions ϕ of a linear operator A. We proceed by the following algorithm:

Step 1. Assume that an approximate eigenfunction of A is of the form $\phi = \phi(x, a)$, where a is a vector of parameters. The form $\phi(x, a)$ must be linear in the second entry a.

Step 2. Place this undetermined $g = \phi$ into the cost functional

$$J = \langle Ag, g\rangle - \lambda\langle g, g\rangle,$$

with λ arbitrary.

Step 3. Search for critical values of J by insisting that the gradient of the cost with respect to the parameters of a is 0, that is,

$$J_a = 0. \tag{6.32}$$

Step 4. Solve the resulting equations between the parameters a and λ.

The resulting λ and corresponding $\phi(x, a)$ are approximate eigenvalues and eigenfunctions of the operator A.

The intuition (and justification) behind this method is an observation by John W. Strutt (Lord Rayleigh) that constraining a vibrating mechanical system will only increase its vibrational frequencies, and as the constraints are gradually removed, the vibrational frequencies will lower asymptotically to the unconstrained frequencies; see [Rayleigh, p. 110]. In Step 1 of the preceding algorithm, we are constraining the vibrating system $\ddot{y} = Ay$ to a finite-degree-of-freedom system; the unconstrained system possesses free vibrational frequencies $\omega = \sqrt{-\lambda}$ corresponding to the *normal modes* ϕ. Consult [Gould] and [Reddy].

Example 4. (CVP 19) Consider the real vector space of all square-integrable realvalued functions f on the interval $[0,1]$. Let us approximate the first several eigenvalues λ and eigenfunctions $\phi = \phi(x)$ of the boundary value problem

$$A\phi = \phi'' = \lambda\phi \tag{6.33}$$

subject to the zero boundary conditions $\phi(0) = 0 = \phi(1)$.

(The exact values are $\lambda_k = -k^2\pi^2$ belonging to $\phi_k = \sin k\pi x$, $k = 1, 2, 3, \ldots$. See Exercise 5.8.)

Note that the cost for this problem is

$$J = \langle A\phi, \phi \rangle - \lambda \langle \phi, \phi \rangle = \int_0^1 \phi''\phi - \lambda\phi^2 \, dx$$

$$= \text{(by parts)} = -\int_0^1 (\phi')^2 + \lambda\phi^2 \, dx. \tag{6.34}$$

Let us try the polynomial

$$\phi = a(x - x^2) + b(x - x^3), \tag{6.35}$$

which certainly satisfies the boundary conditions $\phi(0) = 0 = \phi(1)$. Placing this ϕ into the cost (6.34) yields that

$$-J = \int_0^1 [a(1-2x) + b(1-3x^2)]^2 + \lambda[a(x-x^2) + b(x-x^3)]^2 dx. \tag{6.36}$$

We search for a, b that will yield critical values for J by setting $J_a = J_b = 0$, that is, from (6.36),

$$0 = -\frac{1}{2}\frac{\partial J}{\partial a}$$

$$= \int_0^1 [a(1-2x) + b(1-3x^2)](1-2x) + \lambda[a(x-x^2) + b(x-x^3)](x-x^2) \, dx,$$

6.5 Rayleigh-Ritz Numerics

that is, (Exercise 6.28),

$$0 = 20a + 30b + \lambda(2a + 3b). \tag{6.37a}$$

Likewise,

$$0 = -\frac{1}{2}\frac{\partial J}{\partial b}$$

$$= \int_0^1 [[a(1-2x) + b(1-3x^2)](1-3x^2) + \lambda[a(x-x^2) + b(x-x^3)](x-x^3)\, dx,$$

that is, (Exercise 6.28),

$$0 = 10a + 16b + \lambda(a + 160b/105). \tag{6.37b}$$

Thus we are led to the *generalized eigenvalue problem* $Kv + \lambda Mv = 0$, in this case,

$$\begin{bmatrix} 20 & 30 \\ 10 & 16 \end{bmatrix} \begin{vmatrix} a \\ b \end{vmatrix} + \lambda \begin{bmatrix} 2 & 3 \\ 1 & \frac{160}{105} \end{bmatrix} \begin{vmatrix} a \\ b \end{vmatrix} = 0. \tag{6.38}$$

A MATLAB script for solving such a generalized eigenvalue problem is given in Routine 6.2:

```
% Routine 6.2 generalized eigenvalues
K = [20 30;10 16];         % specify stiffness
M = [2 3; 1 160/105];      % specify mass
[X,D] = eig(K,-M)          % find eigenvectors X, values D
```

The routine returns with

$$X = \begin{bmatrix} 1 & -0.8321 \\ 0 & 0.5547 \end{bmatrix} \text{ and } D = \begin{bmatrix} -10 & 0 \\ 0 & -42 \end{bmatrix}, \tag{6.39}$$

indicating that the estimated two largest eigenvalues are $\lambda_1 = -10$ and $\lambda_2 = -42$ (where the actual values are $-\pi^2 \approx -9.8696$ and $-4\pi^2 \approx -39.4784$) and that their corresponding approximate eigenfunctions are

$$\phi_1 = x - x^2 \text{ and } \phi_2 = -0.8321(x - x^2) + 0.5547(x - x^3). \tag{6.40}$$

These two approximate eigenfunctions are quite close to the actual eigenfunctions (Exercise 6.29).

Example 5. (CVP 20) Find the bounded eigenfunctions ϕ and eigenvalues λ on $-1 \le x \le 1$ of the Legendre operator

$$A\phi = (1-x^2)\phi'' - 2x\phi'. \tag{6.41}$$

Note that the operator A is self-adjoint (Exercise 6.18) on the domain of all twice continuously differentiable functions ϕ on $[-1,1]$ since

$$A\phi = ((1-x^2)\phi')', \tag{6.42}$$

and so, after integrating by parts,

$$\langle A\phi, \psi \rangle = \langle \phi, A\psi \rangle, \tag{6.43}$$

under the natural inner product

$$\langle f, g \rangle = \int_{-1}^{1} f(x)g(x)\,dx. \tag{6.44}$$

We restate this eigenvalue problem as a variational problem: Make stationary the cost

$$J = \langle A\phi, \phi \rangle - \lambda\langle \phi, \phi \rangle = \int_{-1}^{1} ((1-x^2)\phi'(x))'\phi(x) - \lambda\phi^2(x)\,dx$$

$$= -\int_{-1}^{1} (1-x^2)\phi'(x)^2 + \lambda\phi^2(x)\,dx. \tag{6.45}$$

When we search for polynomials that make stationary the integral (6.45) we obtain the useful Legendre polynomials P_n. See Exercise 6.19.

Example 6. (CVP 21) Find the radially symmetric eigenfunctions $\phi = \phi(r)$ and eigenvalues λ of the Laplacian

$$\nabla^2 \phi = \lambda \phi \tag{6.46a}$$

on the annulus

$$\Omega : a < \sqrt{x^2 + y^2} < b \tag{6.46b}$$

subject to zero boundary conditions.

6.5 Rayleigh-Ritz Numerics

The exact analytic solution to this problem is complicated, involving combinations of the Bessel functions J_0 and Y_0; see [MacCluer, 2004]. However, we may restate this problem variationally (Exercise 6.20) as

$$J = \int_\Omega \phi \nabla^2 \phi - \lambda \phi^2 \, d\Omega = \int_0^{2\pi} \int_a^b [\phi(\phi'' + \frac{\phi'}{r}) - \lambda \phi^2] \, r \, dr \, d\theta$$

$$= -2\pi \int_a^b [r(\phi')^2 + \lambda \phi^2 r] \, dr. \tag{6.47}$$

We may now try for approximate solutions numerically using simple polynomial trial functions; see Exercises 6.21–6.22.

Example 7. (CVP 22) To estimate the dominant eigenvalue λ of the problem

$$\nabla^2 \phi = \lambda \phi \quad \text{on } \Omega \quad \text{subject to } \phi = 0 \text{ on } \partial\Omega, \tag{6.48}$$

where Ω is the unit square $0 < x, y < 1$. (The actual value is $\lambda = -2\pi^2 \approx -19.74$.)

We proceed immediately to the variational formulation

$$J = \langle \nabla^2 \phi, \phi \rangle - \lambda \langle \phi, \phi \rangle = \int_\Omega \phi \nabla^2 \phi - \lambda \phi^2 \, d\Omega$$

$$= \text{(Exercise 6.25)} = -\int_\Omega (\nabla \phi)^2 + \lambda \phi^2 \, d\Omega. \tag{6.49}$$

Let us try the polynomial

$$\phi(x) = a(x - x^2)(y - y^2) \tag{6.50}$$

by placing it in (6.49) to obtain when $J_a = 0$ that

$$0 = \int_\Omega (1 - 2x)^2(y - y^2)^2 + (x - x^2)^2(1 - 2y)^2 + \lambda(x - x^2)^2(y - y^2)^2 \, d\Omega$$

$$= 2\left(\int_0^1 (1 - 2x)^2 \, dx\right)\left(\int_0^1 (x - x^2)^2 \, dx\right) + \lambda \left(\int_0^1 (x - x^2)^2 \, dx\right)^2$$

$$= \text{(Exercise 6.26)} = 2 \cdot \frac{1}{3} \cdot \frac{1}{30} + \lambda \left(\frac{1}{30}\right)^2, \tag{6.51}$$

and so $\lambda = -20$.

Further exposition of the Rayleigh-Ritz method can be found in [Strang and Fix].

Exercises

6.1 Prove that for a continuous realvalued function F of a real vector variable x, if

$$\int_\Omega F(x)\eta(x)\, d\Omega = 0$$

for all differentiable η that vanish on the boundary $\partial\Omega$ of the domain $\Omega \subset \mathbb{R}^n$, then $F = 0$ on Ω.

6.2 Provide all the details of a proof of Theorem A.

6.3 Show that every solution of $u_x + u_y = 0$ is of the form $u(x,y) = \phi(x-y)$ for some differentiable ϕ.

Hint: Compute the directional derivative in the direction $(1,1)$.

6.4 Build a nonplanar boundary curve out of wire. Dip this frame into soapy water. Sketch the surfaces assumed by the soap film.

Suggestion: See [Oprea] for a recipe for the best soapy water for such experiments.

6.5 (Project) Prepare a précis of the present status of the double soap bubble problem. What is known and what remains unknown? (The status of this problem is rapidly changing. Search *soap film problem*.)

6.6 Argue that the soap film of Example 2 is a surface of revolution about the z-axis.

6.7 (Transversality) Prove that if the values of $x_k = x_k(u,v)$ are not specified along $\partial\Omega$ (i.e., if the kth component of the surface is unconstrained at its edges), then in addition to (6.8) we have

$$(L_{x_{ku}}, L_{x_{kv}}) \cdot \mathbf{n} = 0 \qquad \text{on } \partial\Omega,$$

where **n** is normal to $\partial\Omega$, or, equivalently,

$$L_{x_{kv}} du - L_{x_{ku}} dv = 0 \qquad \text{along } \partial\Omega.$$

Hint: Apply the divergence theorem to the analogous second integral of (6.4).

6.8 Verify (6.11).

6.9 Consider a surface in $\mathbb{R}^3$ given by

$$x = x(u, v) = (x_1(u, v), x_2(u, v), x_3(u, v)).$$

Argue that the increment of surface area $d\sigma$ is determined by the length of the cross product

$$d\sigma = |x_u \times x_v|\, du\, dv.$$

Compute from (6.8) the analog of (6.11) if this surface is of minimal area.

6.10 Which surfaces of revolution $(x, f(x)\cos\theta, f(x)\sin\theta)$ with $a \le x \le b$, $0 \le \theta \le 2\pi$, are of minimal area?

6.11 [Fox] Find the minimal surfaces of the form $z = f(x) + g(y)$.

Answer:
$$e^{az} = k\cos(ax + c)\,\sec(ay + d).$$

6.12 Reestablish (6.19) via the divergence theorem from the assumption that the net mass flow rate into the region is 0.

6.13 State and prove an analog of (6.6) for an extremal solid $w = w(x, y, z)$.

Answer:
$$L_w - \frac{\partial}{\partial x} L_{w_x} - \frac{\partial}{\partial y} L_{w_y} - \frac{\partial}{\partial z} L_{w_z} = 0.$$

6.14* Consider a distributed structure that at equilibrium occupies the domain Ω. As this structure vibrates, suppose that the displacement u from equilibrium is governed by $u_{tt} = c^2 \nabla^2 u$ with $u = 0$ on $\partial\Omega$. (For example, a vibrating string, drum head, pinned beam, etc.) Show that within units, kinetic energy is given by $T = \langle u_t, u_t \rangle / 2$ and potential energy by $V = -c^2 \langle \nabla^2 u, u \rangle / 2 = c^2 [u, u]/2$, where

$$\langle u, v \rangle = \int_\Omega uv \, d\Omega \quad \text{(Hilbert inner product)}$$

and

$$[u, v] = \int_\Omega \nabla u \cdot \nabla v \, d\Omega \quad \text{(Dirichlet inner product)}.$$

Prove that energy $H = T + V$ is conserved along trajectories.

6.15 Show that self-adjoint matrices [satisfying (6.28) and (6.29)] are exactly the symmetric matrices, that is, matrices that are symmetric about the main diagonal (i.e., $A^T = A$).

6.16 Prove that $A\phi = \phi''$ is a self-adjoint, negative definite operator on all twice-continuously differentiable real-valued functions on $[0, 1]$ satisfying the boundary conditions $\phi(0) = 0 = \phi(1)$.

6.17 Prove that $A\phi = \nabla^2 \phi$ is a self-adjoint, negative definite operator on the set of all twice-continuously differentiable real-valued functions ϕ on the bounded domain $\Omega \subset I\!R^n$ that satisfy $\phi = 0$ on $\partial\Omega$ [under the natural inner product (6.27)].

6.18 Show that the Legendre operator A of Example 5 is indeed self-adjoint.

6.19 Find the first three Legendre polynomials $\phi(x) = a_0 + a_1 x + a_2 x^2$ by finding the singular values of the cost (6.45):

$$F(a_0, a_1, a_2) = \int_{-1}^{1} (1 - x^2) \phi'(x)^2 + \lambda \phi^2(x) \, dx,$$

that is, where

$$\frac{\partial F}{\partial a_0} = \frac{\partial F}{\partial a_1} = \frac{\partial F}{\partial a_2} = 0.$$

Answer: $\phi(x) = 1, x, (3x^2 - 1)/2$, belonging to $\lambda = 0, -2, -6$ respectively.

6.20 Verify (6.47) using the form of the Laplacian in polar coordinates
$$\nabla^2 u = \frac{\partial^2 u}{\partial r^2} + \frac{1}{r}\frac{\partial u}{\partial r} + \frac{1}{r^2}\frac{\partial^2 u}{\partial \theta^2}.$$

6.21 Try for the dominant eigenvalue λ of (6.46) with $a = 1$ and $b = 2$ by placing $\phi(r) = c(r-1)(2-r)$ into the variational formulation (6.47), then proceeding via the Rayleigh-Ritz method as in Example 7.

Answer: $\lambda = -10$.

6.22 Try for the first two eigenvalues and eigenfunctions of (6.46) with $a = 1$ and $b = 2$ by placing $\phi(r) = (r-1)(r-2)(\beta + \alpha r)$ into the variational formulation (6.47), then proceeding via the Rayleigh-Ritz method as in Example 4.

Answer:

λ	α	β
-9.8372	1	-4.157
-42.0983	1	-1.511

Suggestion: Employ *Mathematica* for the integrals and MATLAB for the generalized eigenvalue problem.

6.23 Argue that there exists a scalar field $S = S(x, y, z)$ such that $(\nabla S)^2 = p^2$, making (6.21) possible.

6.24 Prove the converse of Result B: If $f \neq 0$ is critical for the cost (6.30), then λ is an eigenvalue of the self-adjoint operator A belonging to the eigenfunction f, provided that Af belongs to the domain of A (a subspace.)

6.25 Verify (6.49).

6.26 Verify (6.51).

6.27 Show that if $\lambda_1 \leq \lambda_2 \leq \cdots \leq \lambda_k$ are eigenvalues of the linear operator A belonging to the mutually orthogonal eigenvectors $\phi_1, \phi_2, \ldots, \phi_k$, then for any vector η spanned by these eigenvectors,
$$\lambda_1 \langle \eta, \eta \rangle \leq \langle A\eta, \eta \rangle \leq \lambda_k \langle \eta, \eta \rangle.$$
(More generally, *the numerical range of a normal operator is convex and its extreme points are eigenvalues* [MacCluer, 1965].)

6.28 Verify (6.37).

6.29 Using the `plot` and `hold` functions of MATLAB, superimpose the graphs of the approximate eigenfunctions $\phi_1 = x - x^2$ and $\phi_2 = -0.8321(x - x^2) + 0.5547(x - x^3)$ of Example 4 onto the actual eigenfunctions $\psi_1 = \sin \pi x$ and $\psi_2 = \sin 2\pi x$.

6.30 Hand compute the eigenvalues λ and corresponding eigenvectors of
$$A = \begin{bmatrix} 9 & 4 & 0 \\ -6 & -1 & 0 \\ 6 & 4 & 3 \end{bmatrix}.$$
Check your result via MATLAB's `eig`.

6.31 Verify (6.19).

6.32 Verify that (6.25) follows from (6.24).

6.33 (**Dirichlet's Principle**) Assume u is twice-continuously differentiable on some domain containing $\bar{\Omega} = \Omega \cup \partial\Omega$. Prove that if u is a solution of Poisson's equation
$$\nabla^2 u = f \quad \text{on } \Omega$$
subject to $u = g$ on $\partial\Omega$, then among all such twice-continuously differentiable functions satisfying those boundary conditions, u minimizes the cost
$$J = \int_\Omega \nabla u \cdot \nabla u + 2uf \, d\Omega,$$
and conversely.

Outline: (See Exercise 6.14 for the notation.) Consider the perturbed cost $J(\epsilon) = [u + \epsilon\eta, u + \epsilon\eta] + 2\langle f, u + \epsilon\eta \rangle$, where $\eta = 0$ on $\partial\Omega$. Then $J(\epsilon) = J(0) + \epsilon J'(0) + \epsilon^2[\eta, \eta]$. But $\nabla^2 u = f$ implies $J'(0) = 0$ and so $J(\epsilon) \geq J(0)$. Conversely, any minimizer u satisfies the Lagrange equation (i.e., $\nabla^2 u = f$).

Chapter 7
Optimal Control

There are two schools of thought about the theory of control—*open-loop optimal control* and *closed-loop feedback control*. Both have enjoyed remarkable successes. This chapter is a short introduction to the open-loop school. We begin with several attempts to control a rolling cart, motivating a general formulation of optimal control problems. We then work problems on reinvestments and average voltages in an *RC* circuit. A simple time-optimal problem introduces the "bang-bang" principle. We conclude with the important maximum principle of Pontryagin and example applications.

7.1 A Rolling Cart (OCP 1)

Let us begin with a traditional first example of this theory. Our *task* will be to move the cart of mass $m = 1$ shown in Figure 7.1 from $x = 0$ to $x = 100$ in exactly $T = 10$ seconds by applying a nonnegative force u.

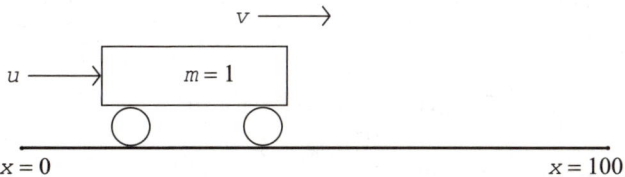

Figure 7.1 Our task is to move the cart a distance of 100 with the least amount of work *W*.

Our *goal* is to do this task with the least work possible starting at rest $x(0) = v(0) = 0$. But work W is the time-integral of the product of force u with velocity $v = \dot{x}$, that is,

$$W = \int_0^{100} u \, dx = \int_0^{10} uv \, dt. \tag{7.1}$$

Let us ignore the complicating factors of rolling friction, air resistance, and the rotational inertia of the wheels. Balancing inertial force by the single external force u we obtain the dynamic model

$$\dot{v} = u, \tag{7.2}$$

and so

$$W = \int_0^{10} \dot{v} v \, dt = \frac{v(10)^2}{2}. \tag{7.3}$$

Solution Attempt 1. Our objective is to minimize the work W of (7.3) subject to the integral constraint

$$\int_0^{10} v \, dt = x(10) - x(0) = 100 \tag{7.4}$$

with a nonnegative control[1] subject to the end condition $v(0) = 0$.

Impose the Euler-Lagrange equation on the augmented Lagrangian $N = \dot{v}v + \lambda v$. No information is revealed about the *optimal control force* u that accomplishes the task with minimal work (Exercise 7.3). This is because the very best nonnegative control strategy (were it physically possible) would be an impulse, an instantaneous hammer blow at $t = 0$, yielding total work $W = 50$. See Exercises 7.1–7.4.

Solution Attempt 2. To disallow such impulsive, non-physical control signals, let us penalize large forces with a modified cost

$$J = \int_0^{10} \epsilon u^2 + uv \, dt, \tag{7.5}$$

subject to the integral constraint (7.4). Intuitively, when $\epsilon > 0$ is small, the cost J should be essentially the work W.

Imposing the Euler-Lagrange equations on the augmented Lagrangian $N = \epsilon \dot{v}^2 + \dot{v}v + \lambda v$ yields (Exercise 7.13)

$$\ddot{v} = \lambda/2\epsilon, \tag{7.6}$$

and thus the control force is linear:

$$u(t) = \lambda t/2\epsilon + u(0). \tag{7.7}$$

---

[1] It is easy to see (Exercise 7.1) that $W \geq 50$ (in the appropriate units) when only nonnegative force u is allowed. If we allow negative (retarding) forces, we can accomplish the task with zero net work; see Exercise 7.2.

7.1 A Rolling Cart

If $u \equiv u(0)$, then $v = u(0)t$ and $x = u(0)t^2/2$, giving that $u(0) = 2$ since $x(10) = 100$. The total work done by such a constant force is $W = v(10)^2/2 = 200$.

More generally, any linear nonnegative control (7.7) that accomplishes the goal will do work $W \geq 112.5$. This lower bound is achieved with the control $u = -3t/10 + 3$ (Exercise 7.5).

Unfortunately, by transversality (Exercise 7.31), since the final velocity $v(10)$ is unspecified, $0 = N_{\dot{v}}|_{t=10} = 2\epsilon\dot{v}(10) + v(10)$, which shows no minimization of (7.5) is possible with a nonnegative control force u.

Solution Attempt 3. Let us severely penalize the use of unnecessary work by means of the cost

$$J = \int_0^{10} (uv)^2 \, dt = \int_0^{10} \dot{v}^2 v^2 \, dt, \tag{7.8}$$

subject to the end conditions $x(0) = v(0) = 0$ and the integral constraint (7.4).

Apply the Euler-Lagrange equation to the augmented Lagrangian

$$N = \dot{v}^2 v^2 + \lambda v \tag{7.9}$$

to obtain that $\lambda = 0$ [because $v(0) = 0$] and that when $v\dot{v} \neq 0$ (Exercise 7.6):

$$-\dot{v}/v = \ddot{v}/\dot{v}, \tag{7.10}$$

$$2v\dot{v} = k^2, \tag{7.11}$$

$$v = k\sqrt{t}, \tag{7.12}$$

$$x = 2kt^{3/2}/3, \tag{7.13}$$

giving from $x(10) = 100$ that $k = 15/\sqrt{10}$. Hence total work done under this control strategy is

$$W = v(10)^2/2 = (k\sqrt{10})^2/2 = 225/2 = 112.5, \tag{7.14}$$

no improvement over our previous attempt. Moreover, our control force

$$u = \dot{v} = k/2\sqrt{t}, \tag{7.15}$$

although positive, is unbounded at $t = 0$. Note that transversality requires (Exercise 7.14) that $\dot{v}(10)v(10) = 0$ and so (7.15) cannot be optimal for the cost (7.8).

Solution Attempt 4. To disallow unbounded control forces u, let us try the marginally related cost

$$J = \int_0^{10} u^2 + v^2 \, dt = \int_0^{10} \dot{v}^2 + v^2 \, dt. \tag{7.16}$$

Applying the Euler-Lagrange equation to the augmented Lagrangian

$$N = \dot{v}^2 + v^2 + 2\lambda v, \tag{7.17}$$

we obtain
$$\ddot{v} - v = \lambda. \tag{7.18}$$

Solving (7.18) and imposing $x(0) = v(0) = 0$, $x(10) = 100$, and transversality $\dot{v}(10) = 0$ yields (Exercise 7.8)

$$v = \left[\frac{100}{10 - \tanh 10} \right] \left[1 - \frac{\cosh(10-t)}{\cosh 10} \right]. \tag{7.19}$$

The work done by the resulting nonnegative bounded control $u = \dot{v}$ is the much improved $W = v(10)^2/2 = 61.7$.

7.2 General Formulation

Optimal control theory is the search for a *optimal control signal* $u = u(t)$ among a class U of *admissible* controls that will steer a system governed by

$$\dot{x} = F(t, x, u) \tag{7.20a}$$

from a given *initial state* x_0 to a given *target state* x_1, and, while doing so, minimizes a given *performance index (cost functional)*

$$J = \int_a^b L(t, x, u) \, dt. \tag{7.20b}$$

Theorem A. Suppose both F and L are continuously differentiable. If a continuous control signal $u = u(t)$ steers the system (7.20a) to the minimal cost J of (7.20b), then it is necessary, for some piecewise continuously differentiable $\lambda = \lambda(t)$, that

$$N = L(t, x, u) + \lambda(t)(\dot{x} - F(t, x, u)) \tag{7.21}$$

satisfies the Lagrange equations.

Proof. Theorem D, §5.7.

7.3 Reinvestments (OCP 2)

In most applications, the state x is a vector variable and so the product with λ in (7.21) is a dot product. See Exercise 7.12 and Figure 7.2.

Critique. The rolling cart problem of Figure 7.1 captures many typical characteristics of optimal control theory:

- The optimal control u is often impulsive, or discontinuous pulses. Thus a naive application of Theorem A will rarely succeed.

- The performance index must often be modified, resulting in a cost functional having marginal physical or economic meaning. For example, the quadratic cost J of (7.8) or (7.16) has only slight relevance to the actual work W of (7.1).

- The parameters of a derived optimal control signal u are often not robust to perturbations—small changes in the control signal u may result in large deviations in the performance indices (see Exercise 7.22).

- The application of optimal control to practical problems is an art, requiring the practitioner to perform many analytic and numerical iterations to reach an acceptable (but often not optimal) solution to the original control problem.

Let us then obtain some practice with this art.

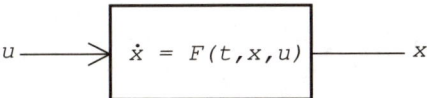

Figure 7.2 An open-loop control system with control signal u and state variable x.

7.3 Reinvestments (OCP 2)

OCP 2. What portion of output should be reinvested to maximize marketable production?

We are producing a commodity at the rate of $q = q(t)$ units per unit time t. But in the first approximation, the rate $\dot{q}$ of production increase is proportional to the percentage u of the production q reinvested. Thus our goal is to maximize total marketable production

$$Q = \int_0^T (1-u)q \, dt \qquad (7.22a)$$

during the time period $[0, T]$ subject to

$$\dot{q} = \alpha u q. \qquad (7.22b)$$

But imposing the Lagrange equations upon

$$N = (1-u)q + \lambda(t)(\dot{q} - \alpha u q) \qquad (7.23)$$

will once again yield no useful information about any possible (differentiable) control u (Exercise 7.15).

So once more let us turn to a quadratic performance measure:

$$J = \int_0^T (1-u)^2 q^2 \, dt \qquad (7.24a)$$

subject to

$$\dot{q} = \alpha u q. \qquad (7.24b)$$

Imposing the Euler-Lagrange equations on

$$N = (1-u)^2 q^2 + \lambda(t)(\dot{q} - \alpha u q) \qquad (7.25)$$

will now yield useful insight (Exercise 7.16):

$$\dot{u} = \alpha(1 - u^2), \qquad (7.26)$$

and hence by partial fractions,

$$u = \frac{ke^{2\alpha t} - 1}{ke^{2\alpha t} + 1}, \quad k > 1. \qquad (7.27)$$

Thus the total production for the period $[0, T]$ is

$$q = K \exp[\alpha \int_0^t \frac{ke^{2\alpha \tau} - 1}{ke^{2\alpha \tau} + 1} \, d\tau], \qquad (7.28)$$

and so the control signal u of (7.27) results in an eventual exponential growth in production rate:

$$q \approx K e^{\alpha t},$$

7.4 Average Voltage (OCP 3)

even though an increasing percentage of production is being reinvested.

7.4 Average Voltage (OCP 3)

Apply the control voltage u to the input of the low-pass filter shown in Figure 7.3 [Pierre].

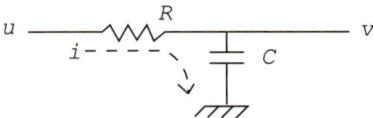

Figure 7.3 An *RC* lowpass filter.

We wish to maximize the average output voltage

$$A = \frac{1}{T} \int_0^T v \, dt$$

during the time period $0 \leq t \leq T$ by a clever choice of u without exceeding the energy dissipation E of the resistor, that is, where

$$\int_0^T i^2 R \, dt \leq E.$$

Intuition tells us that if we begin with a high input voltage u, the capacitor C will draw a large current, thus increasing the drop across R, thus decreasing the voltage v across C.

Recall that the relation between input and output (assuming no current is drawn from the output terminal) is

$$RC\dot{v} + v = u$$

and the current i through the capacitor is

$$i = C\dot{v},$$

see [MacCluer, 2000]. Thus our variational task is to maximize

$$A = \frac{1}{T} \int_0^T v \, dt \qquad (7.29a)$$

subject to the integral constraint

$$\int_0^T RC^2 \dot{v}^2 \, dt = E, \qquad (7.29\text{b})$$

and the condition (say)
$$v(0) = 0. \qquad (7.29\text{c})$$

To illustrate the ideas let us do the case $R = C = T = 1$. Imposing the Euler-Lagrange equation upon the augmented Lagrangian

$$N = v + \lambda \dot{v}^2$$

yields (Exercise 7.17) that

$$v = a + bt + ct^2. \qquad (7.30)$$

Hence the proper control signal $u = \dot{v} + v$ is also quadratic (see Exercise 7.18).

7.5 A Time-Optimal Problem (OCP 4)

A large class of important control problems involve steering a system to a specified final state in the least time.

The Rolling Cart Reprised (OCP 4). Let us return to our cart of Figure 7.1, but now assume the cart has initial displacement $x(0) = x_0$ and velocity $v(0) = v_0$. Our task is a docking problem—to return the cart to $0 = x(T)$ and arrive with terminal velocity $0 = v(T)$ using a *bang-bang* control strategy,[2] where the force u is either full on to the right $u = 1$ or full reverse to the left $u = -1$. Our goal will be to accomplish this task in the least time T. Can we accomplish all this by a clever switching strategy?

The task is possible: We can return the cart to $x(T) = 0$ and arrive with velocity $v(T) = 0$ in some finite time T by the following graphical strategy: During the intervals when $u = 1$, $v = t + a$ and so by completing the square, $x = t^2/2 + at + b = (t+a)^2/2 + b - a^2/2 = v^2/2 + \beta$. Thus the cart is traveling in phase space (x, v) along a parabola opening upward to the right as in Figure 7.4.

---
[2] A *bang-bang* control signal $u(t)$ takes on only two distinct values: a or $-a$.

7.5 A Time-Optimal Problem (OCP 4)

On the other hand, when the thrust is in reverse, $u = -1$, we see likewise that $x = -v^2/2 + \gamma$, and thus the trajectories are downward along parabolas opening to the left (again see Figure 7.4). So the switching strategy is clear:

Choose a path along intersecting parabolic segments until reaching one of the two parabolas that lead to the origin. At most one switching is necessary.

For example, as in Figure 7.4, starting with the initial conditions (x_0, v_0) at a, follow the downward parabola ($u = -1$) to b, continue on until reaching c, then return to the origin along $2x = v^2$.

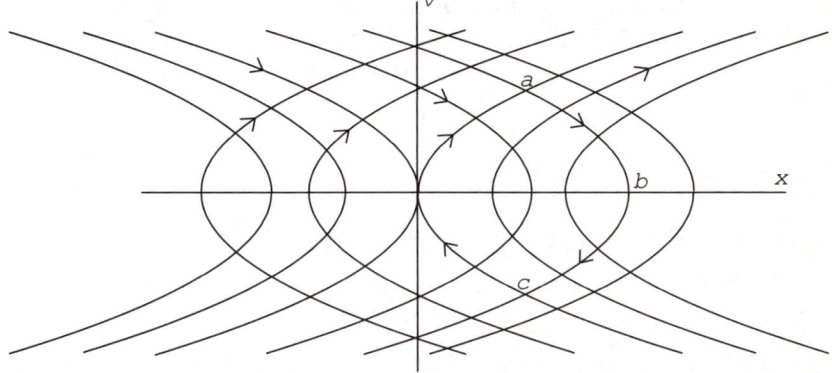

Figure 7.4 The minimal time to return to (0,0) in phase space is achieved by following parabolic segments. For example, starting at a, take the path to b, continue on to c, then switch to the parabola through (0,0).

Have you guessed the truth? This strategy is optimal! *There is no other control $u = u(t)$ with $|u(t)| \leq 1$ (continuous or otherwise) that achieves the task is less time.*

We can see that the bang-bang control is time optimal in several ways. If (x, v) is heading away from the origin, then the tightest turn back toward the origin is with opposing acceleration full on, $|u| = 1$. Any less than this acceleration will leave the point (x, v) outside of the two-piece bang-bang parabolic paths of Figure 7.4. More analytically, the time ΔT to traverse a path $(x(t), v(t))$ leading from (x_0, v_0) to the origin is the line integral

$$\Delta T = \int_{(x_0, v_0)}^{(0,0)} \frac{dx}{v} = \int_{(x_0, v_0)}^{(0,0)} \frac{|dx|}{|v|}$$

along the path.[3] But after parametrizing by

$$s = \int_{(x_0,v_0)}^{(x,v)} |dx|,$$

and noting that

$$u = \dot{v} = \frac{dv}{ds}\frac{ds}{dt} = v'|v|,$$

$$\Delta T = \int_0^s \frac{ds}{|v|} = \int_0^s \frac{1}{|u|}|v'|\,ds \geq \int_0^s |v'|\,ds = \sum_i |v_{i+1} - v_i|, \quad (7.31)$$

where the sum is over the intervals where v is monotonic. But this inequality is in fact an equality along our bang-bang parabolic paths $2x = \pm v^2 + c$. This leads to a proof that the one-switch scheme is the time-optimal control strategy (Exercise 7.23).

The preceding example is an illustration of a general rule.

7.6 The Bang-Bang Principle

Suppose we are tasked with steering a time-invariant linear system governed by

$$\dot{x} = Ax + Bu, \quad (7.32)$$

where the state x and the control u are $n \times 1$ and $r \times 1$ columns, respectively. Thus the (constant) matrices A and B are of size $n \times n$ and $n \times r$, respectively.

Recall that the trajectory of the system (7.32) initiating at state $x(0) = x^0$ is

$$x(t) = e^{At}x^0 + \int_0^t e^{A(t-\tau)}Bu(\tau)\,d\tau. \quad (7.33)$$

See Exercise 7.24. There is this startling result.

Theorem B. (bang-bang)[4] If a bounded control $u^0 = u^0(t)$ steers a time-invariant linear system $\dot{x} = Ax + Bu$ from the initial state $x^0 = x(0)$ to a target state x^1 in time T, then this task can also be achieved by a bang-bang control u in time T.

---

[3] Surely we need not consider strategies where the cart remains at rest for any length of time.

[4] A control u is *bang-bang* if each of its coordinate functions $u_i(t)$ are bang-bang.

7.6 The Bang-Bang Principle

Proof Sketch.* By rescaling each column of the *input matrix B*, we may assume
$$|u^0|_\infty = 1,$$
that is, each coordinate function of u^0 has 1 as its maximum absolute value on $[0, T]$. Consider the set U of all bounded controls $u = u(t)$ on $0 \le t \le T$ with $|u|_\infty \le 1$ that steer the system from $x^0 = x(0)$ to state $x^1 = x(T)$ in time T. It is clear that U is a convex set containing at least the control u^0.

By one of the pillars of functional analysis, the Banach-Aloglu theorem, the *closed unit ball* $\mathcal{B}$ of all measurable r-vector functions u on the interval $[0, T]$ with $|u|_\infty \le 1$ is a compact set under the weak* topology [Reed and Simon]. It is easy to see that the set of admissible controls U is weak*-closed and hence weak*-compact (Exercise 7.25). But by Exercise 7.26, compact convex sets possess extreme points. Finally, by Exercises 7.27 and 7.28, extreme controls are bang-bang.

Remark. Theorem B can be extended to time-varying linear systems $\dot{x} = A(t)x + B(t)u$ for locally integrable $A(t)$ and $B(t)$—see the wonderful book by G. Knowles. Lee Sonneborn outrageously claims the bang-bang theorem reveals that one can drive to the local convenience store by cleverly alternating between full acceleration and full braking, and probably in less time than usual.

Corollary A. The bang-bang control u whose existence is guaranteed by the theorem has the same bound as the control u^0 it replaces. That is, if $\dot{x} = Ax + Bu$ can be steered from state $x^0 = x(0)$ to state x^1 with the bounded control u, then it can be accomplished with a bang-bang time-optimal control $|u|_\infty = |u^0|_\infty$ (Exercise 7.41).[5]

Corollary B. If the system $\dot{x} = Ax + Bu$ is *controllable*, then given any prescribed steering task, there is a time-optimal bang-bang control that can accomplish it. (See Exercise 7.42.)

---

[5]The notation $|u^1|_\infty = |u^2|_\infty$ means the respective coordinate functions of u^1 and u^2 are in that relation.

7.7 The Maximum Principle

We have now seen in several examples that the Lagrange equations are not always successful in revealing an optimal control, often because such controls are not even continuous, much less piecewise continuously differentiable. For instance, Theorem B demonstrates that optimal bounded controls are often bang-bang. We need a second tool to discover these hidden discontinuous controls.

Theorem C. (Pontryagin, ca. 1959) The admissible controls U are to be all piecewise continuous[6] functions u whose values $u(t)$ lie in some prescribed compact set C. Suppose there exists an optimal control $u = u^*(t)$ of U that steers the (time-invariant) system

$$\dot{x} = F(x, u) \qquad (7.34a)$$

along the curve $x = x^*(t)$ in such a way that minimizes the cost[7]

$$J = \int_0^T L(x, u)\, dt \qquad (7.34b)$$

while satisfying the end conditions $x(0) = a$ and $x(T) = b$ for a specified T. Assume that L and F are everywhere continuously differentiable. Define the *Hamiltonian*

$$H(x, \lambda, u) = -L(x, u) + \lambda \cdot F(x, u), \qquad (7.35a)$$

where λ is called the *adjoint variable*. Then there is a (piecewise continuously differentiable) solution $\lambda = \lambda(t)$ of the *adjoint equation*

$$\dot{\lambda} = -H_x(x^*, \lambda, u^*) \qquad (7.35b)$$

for which the resulting Hamiltonian H is constant along the optimal trajectory $(x^*(t), \lambda(t), u^*(t))$ for $0 \leq t \leq T$, and this constant value is given by

$$E = \sup_v H(x^*(t), \lambda(t), v), \qquad (7.36)$$

where the supremum is taken over all $v \in C$.

---

[6] A function is *piecewise continuous* if it possesses at most a finite number of discontinuities, all either removable or jump discontinuities.

[7] We tacitly assume that we are not dealing with an *abnormal* system in which the cost is unaffected by the choice of path taken.

7.7 The Maximum Principle

Proof under additional hypotheses.[8] (All vectors are to be row vectors.) Because the controls u are piecewise continuous, by Picard's theorem (Appendix B), all resulting paths $x = x(t)$ satisfying (7.34a) are piecewise continuously differentiable. Hence solutions $\lambda = \lambda(t)$ to (7.35b) must also be piecewise continuously differentiable (Exercise 7.43).

Note that if u is optimal on $[t_1, t_2]$, it is optimal on every subinterval (Exercise 7.44).

Assume that for *each* initial point $x(0) = x^0$, there is an optimal control $u = u(x^0, t)$ that determines an optimal path $x = x(x^0, t)$ from x^0 to b in some finite time T_0 that minimizes the cost

$$J(x^0) = \int_0^{T_0} L(x, u) \, dt.$$

Note by optimality that

$$J(x(x^0, t)) = \int_t^{T_0} L(x(x^0, s), u(x^0, s)) \, ds.$$

Assuming $J(x)$ is differentiable and the controls are piecewise continuous, we may differentiate the above identity with respect to t to see that

$$0 = L(x(x^0, t), u(x^0, t)) + \nabla J(x(x^0, t)) \cdot \dot{x}(x^0, t)$$
$$= L(x(x^0, t), u(x^0, t)) + \nabla J(x(x^0, t)) \cdot F(x(x^0, t), u(x^0, t)) \quad (7.37a)$$

for all but a finite number (a.b.f.n.) of t and all x^0. In particular, when $u = u^*(t)$, $x(t) = x^*(t)$, and $x^0 = a$,

$$0 = L(x^*, u^*) - \lambda \cdot F(x^*, u^*) = -H(x^*, \lambda, u^*) \quad (7.37b)$$

for a.e. $0 \le t \le T$, where

$$\lambda(t) = -\nabla J(x(a, t)).$$

Thus H is constantly 0 a.b.f.n. along the optimal path (x^*, λ, u^*) for this choice of λ.

---

[8] From [Macki and Strauss], page 120.

For any constant $v \in C$, solve $\dot{x} = F(x,v)$ for a local solution $x = \tilde{x}(t)$ that initiates at $x^0 = \tilde{x}(0)$. Since in general $x = \tilde{x}$ is not the optimal path from x^0 to $\tilde{x}(t)$, it follows that

$$J(x^0) \le \int_0^t L(\tilde{x}, v) \, ds + J(\tilde{x}(t)),$$

that is, for $t > 0$,

$$-\frac{J(\tilde{x}(t)) - J(x^0)}{t} \le \frac{1}{t} \int_0^t L(\tilde{x}, v) \, ds,$$

and so in the limit, since $L(\tilde{x}, v)$ is continuous in t,

$$-\frac{d}{dt} J(\tilde{x}(t)) \bigg|_{t=0} = -\nabla J(x^0) \cdot \dot{\tilde{x}}(0)$$

$$= -\nabla J(x^0) \cdot F(x^0, v) \le L(x^0, v). \tag{7.38a}$$

In particular, taking x^0 to be any point along an optimal path $x = x^*(t)$, we have

$$H(x^*(t), \lambda(t), v) \le 0 \tag{7.38b}$$

for a.b.f.n. $0 \le t \le T$ and all $v \in C$.

As a consequence of (7.38a) and (7.37b), the function

$$h(x, u) = -L(x, u) - \nabla J(x) \cdot F(x, u)$$

is maximum when $x = x^*(t)$ and $u = u^*(t)$ at a.b.f.n. t. Assuming $J(x)$ is twice continuously differentiable at each x, we must then have along $x = x^*(t)$ that (Exercise 7.46)

$$0 = \frac{\partial h}{\partial x_i} = -\frac{\partial L}{\partial x_i} - \frac{\partial (\nabla J \cdot F)}{\partial x_i}$$

$$= -\frac{\partial L}{\partial x_i} - \frac{\partial}{\partial x_i} \sum_{j=1}^n \frac{\partial J}{\partial x_j} F_j = -\frac{\partial L}{\partial x_i} - \sum_{j=1}^n \frac{\partial^2 J}{\partial x_i \partial x_j} F_j - \sum_{j=1}^n \frac{\partial J}{\partial x_j} \frac{\partial F_j}{\partial x_i}$$

$$= -\frac{\partial L}{\partial x_i} + \sum_{j=1}^n \frac{\partial \lambda_i}{\partial x_j} \dot{x}_j + \sum_{j=1}^n \lambda_j \frac{\partial F_j}{\partial x_i}. \tag{7.39}$$

Hence

$$\dot{\lambda}_i = -\lambda \cdot F_{x_i} + L_{x_i},$$

that is, λ satisfies the adjoint equation (7.34b).

7.7 The Maximum Principle

Remark. The conclusions of Pontryagin's maximal principle remain true when the admissible controls U are changed from piecewise continuous to measurable (with values in C). However, x and λ become merely absolutely continuous with bounded derivative. See [Macki and Strauss].

Corollary. If $H(x, \lambda, u)$ is convex in u and the admissible controls are $U = \mathcal{B}$ (i.e., all $|u|_\infty \leq 1$), then the optimal control is a bang-bang control. In particular, if H is *affine* in u [i.e., of the form $H(x, \lambda, u) = A(x, \lambda)u + B(x, \lambda)$], then the optimal control is bang-bang under some mild additional assumptions. See [Bell and Jacobson].

Proof Sketch.* By Bauer's maximum principle (Exercise 7.29), maxima of convex functions on compact convex sets occur at extreme points. But by Exercise 7.27, the extreme points of the set $\mathcal{B}$ of all $|u(t)|_\infty \leq 1$ are exactly the bang-bang controls.

Remark. An often useful variant on the maximum principle is where the time T to target is specified but where the target value $b = x^*(T)$ is not. The preceding proof can be modified to prove this variant (Exercise 7.50). Moreover, because $b = x^*(T)$ is not specified, the adjoint variable must terminate with the value 0, that is, we have the *transversality* result (Exercise 7.51)

$$\lambda(T) = 0.$$

More delicately, if, say, the ith coordinate b_i of the target is unspecified, the corresponding coordinate $\lambda_i(T) = 0$. As we shall now see, this valuable additional information can be used to determine control signal switching times.

Alert. The maximal principle establishes *necessary* conditions on the optimal control signal u. There is no guarantee that the system possesses an optimal control at all.

7.8 Example Applications

Let us look at previous problems where the Euler-Lagrange equations failed to reveal the optimal control but where the maximum principle succeeds.

Example 1. (OCP 2 revisited) What portion of output should be reinvested to maximize marketable production?

We are to maximize

$$Q = \int_0^T (1-u)q \, dt \qquad (7.40a)$$

subject to

$$\dot{q} = \alpha u q. \qquad (7.40b)$$

The admissible controls U are to be piecewise continuous where of course $0 \leq u \leq 1$.

Step 1. Find the Hamiltonian and deduce the optimal control.

Since we are to minimize $-Q$, the Hamiltonian $H(q, \lambda, u)$ is

$$H = -L + \lambda F = (1-u)q + \alpha \lambda u q = (1 - u + \alpha \lambda u)q. \qquad (7.41)$$

Fix q and λ as arising from the optimal solution u. At each time t, because $q > 0$, H is maximized exactly when $-u + \alpha \lambda u = u(\alpha \lambda - 1)$ is maximized. But since $0 \leq u \leq 1$, the optimal control is

$$u(t) = 1 \quad \text{if} \quad \alpha\lambda(t) > 1 \qquad \text{and} \qquad u(t) = 0 \quad \text{if} \quad \alpha\lambda(t) < 1.$$

Thus the optimal control is two-valued, switching when $\alpha\lambda - 1$ changes sign.

Step 2. Solve the adjoint equation.

The parameter λ satisfies the *adjoint* equation (7.38), which in this case is $\dot{\lambda} = -H_q = u - 1 - \alpha \lambda u$, that is,

$$\dot{\lambda} + \alpha \lambda u = u - 1, \qquad (7.42)$$

which become the differential equations

$$\dot{\lambda} + \alpha\lambda = 0 \quad \text{when} \quad u(t) = 1 \qquad \text{and} \qquad \dot{\lambda} = -1 \quad \text{when} \quad u(t) = 0,$$

7.8 Example Applications

with solutions

$$\lambda = ce^{-\alpha t} \qquad \text{when } u(t) = 1, \qquad (7.43a)$$

and

$$\lambda = -t + d \qquad \text{when } u(t) = 0. \qquad (7.43b)$$

Step 3. Specify the end condition $\lambda(T)$.

Since $q(T)$ is unspecified, by the transversality condition (Exercise 7.51), $\lambda(T) = 0$. Thus (7.43a) cannot obtain when $t = T$. Thus (7.43b) must obtain and $d = T$, in short,

> *toward the end of the time period, reinvest no production.*

Step 4. Deduce the switching pattern.

Working backward in time from $t = T$, we see that $\alpha\lambda(t) = \alpha(T-t) < 1$ until $t_s = T - 1/\alpha$, where we switch to total reinvestment $u = 1$. So in short, the optimal strategy (if it exists) must be to

> *at first reinvest all production until $t = T - 1/\alpha$,*
> *thereafter sell all goods produced.*

Example 2. (OCP 1 revisited) The rolling cart of Figure 7.1.

Starting from rest at $x = 0$, we are to command the cart to arrive at $x = 100$ in $T = 10$ seconds by applying the force u. We search for a piecewise continuous bounded control $0 \le u(t) \le a$ that will accomplish this task with least work

$$W = \int_0^{10} uv \, dt, \qquad (7.44)$$

where the dynamics are given by $(\dot{x}, \dot{v}) = F(x, v) = (v, u)$.

The Hamiltonian is

$$H = -L + \lambda F = -uv + (\alpha, \beta) \cdot (v, u) = -uv + \alpha v + \beta u, \qquad (7.45)$$

where the adjoint variable $\lambda = (\alpha, \beta)$. The Hamiltonian H along the optimal trajectory (x, v) is maximized exactly when $u(\beta - v)$ is maximized, and hence by the maximum principle, because $0 \le u(t) \le a$, the optimal control u can take on only the values $u(t) = a$ or $u(t) = 0$ according to whether $\beta - v > 0$ or $\beta - v < 0$, respectively.

During those same periods, since $\dot{v} = u$, $v = at + b$ or $v = v_0$, respectively.

On the other hand, the adjoint equation

$$\dot{\lambda} = (\dot{\alpha}, \dot{\beta}) = -H_{(x,v)} = -(0, -u + \alpha) = (0, u - \alpha)$$

has solutions of the form $\alpha = \alpha_0$ and hence

$$\beta = \begin{cases} -(\alpha_0 - a)t + c & \text{when } u = a \\ -\alpha_0 t + d & \text{when } u = 0 \end{cases}. \quad (7.46)$$

Thus in either case we have

$$\beta - v = -\alpha_0 t + \text{ constant}.$$

But both v and β must be continuous. Hence for some c_0 and all $0 \le t \le 10$,

$$\beta - v = -\alpha_0 t + c_0. \quad (7.47)$$

Thus there can be at most one switching (i.e., one zero crossing of $\beta - v$).

Moreover, from transversality (Exercise 7.51), since the terminal velocity $v(10)$ is unspecified, $\beta(10) = 0$. Finally, since $u \ge 0$, $v(10) > 0$. Hence $\beta - v$ is negative at $t = 10$ and thus the control signal $u = 0$ as the cart crosses the finish line at $t = 10$. But this implies there has been exactly one switching as in Figure 7.5, otherwise $u \equiv 0$ and the cart never leaves its initial position at $x = 0$.

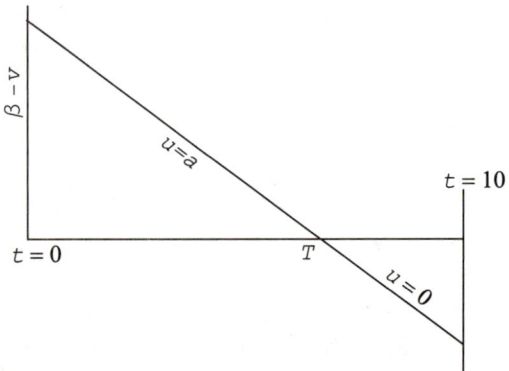

Figure 7.5 The graph of $\beta - v$. The optimal control must switch from $u = a$ to $u = 0$ at the moment that $\beta - v$ becomes negative.

We now choose on-time T to reach $x = 100$ when $t = 10$ (Exercise 7.34).

7.8 Example Applications

Example 3. (OCP 4) Consider Bushaw's example

$$\dot{x} = y, \quad \dot{y} = -x + u, \quad |u| \leq 1. \tag{7.48}$$

Our task is to find the control u that steers this system from a given initial state (x_0, y_0) to the origin in the least time T. The equations describe the displacement x and velocity y of a sliding mass $m = 1$ with a restoring spring of spring-constant $k = 1$. See Figure 7.6.

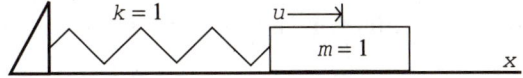

Figure 7.6 A sliding mass with a restoring spring.

If this task can be accomplished, it can be done optimally with a bang-bang control since this is a linear time-optimal problem. Let us apply the Pontryagin maximum principle to actually discover this optimal strategy.

Since $u = \pm 1$, between times of switching the states satisfy $\ddot{x} + x = u$ with solution

$$x = a\cos(t - \phi) + u \tag{7.49a}$$

and hence

$$y = -a\sin(t - \phi). \tag{7.49b}$$

Thus between switchings, the states (x, y) are orbiting clockwise along either the circle

$$(x - 1)^2 + y^2 = a^2 \quad \text{when } u = 1 \tag{7.50a}$$

or the circle

$$(x + 1)^2 + y^2 = a^2 \quad \text{when } u = -1; \tag{7.50b}$$

see Figure 7.7.

Consider the Hamiltonian of this problem:

$$H = -L + \lambda \cdot F = -1 + (\alpha, \beta) \cdot (y, -x + u) = -1 + \alpha y - \beta x + \beta u. \tag{7.51}$$

Note that H maximizes exactly when βu maximizes and so

$$u(t) = \operatorname{sgn} \beta(t), \tag{7.52}$$

independently establishing that the optimal control is indeed bang-bang (see Exercise 7.30) where switching occurs as β changes signs. But by the adjoint equation $\dot{\lambda} = -H_{(x,y)}$, we see that $\dot{\alpha} = -\beta$, $\dot{\beta} = \alpha$, hence $\ddot{\beta} = -\beta$, giving

$$\beta(t) = b\sin(t - \psi). \tag{7.53}$$

Thus between the first and last,

switching occurs every π seconds!

This amounts to reflections through either $(-1, 0)$ or $(1, 0)$.

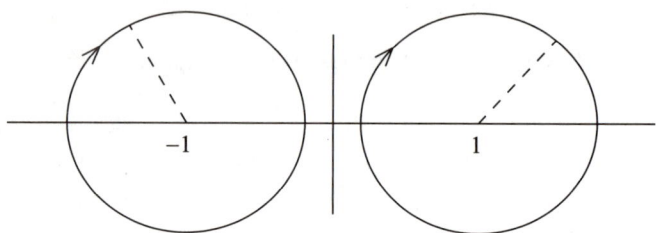

Figure 7.7 Between switchings, the system orbits clockwise half way around a circle of radius *a* centered at (-1, 0) if *u* = –1, or at (1, 0) if *u* = 1.

Because this system is controllable (Exercise 7.36), a unique time-optimal (bang-bang) control exists. Here is how it is found:

Consider the unit circles $(x - m)^2 + y^2 = 1$ with m an odd integer as in Figure 7.8. If we start with a point on one of the lower semi-circles in the right half-plane and begin with thrust to the right (i.e., $u = 1$), the trajectory will in steps of $t = \pi$ seconds, move toward the origin on such circles, eventually hitting one of the critical unit circles with center $(-1, 0)$ or $(1, 0)$, whereupon it can lead to $(0, 0)$ through an angle less than π. See Figure 7.8.

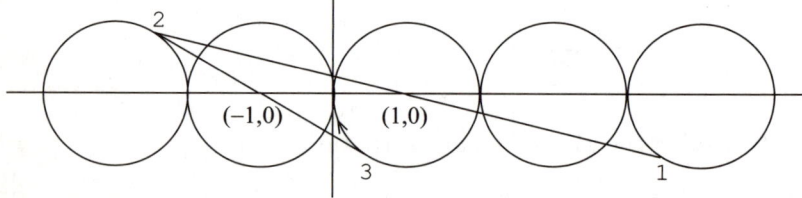

Figure 7.8 Starting at a point on any one of the unit circles with center *m*, *m* odd, subsequent points of switching also lie on such circles but closer to (0,0). Eventually one of the two critical circles with center (−1, 0) or (1, 0) is reached.

Exercises

The same result holds when we start from the upper semicircles in the left-hand plane with thrust to the left.

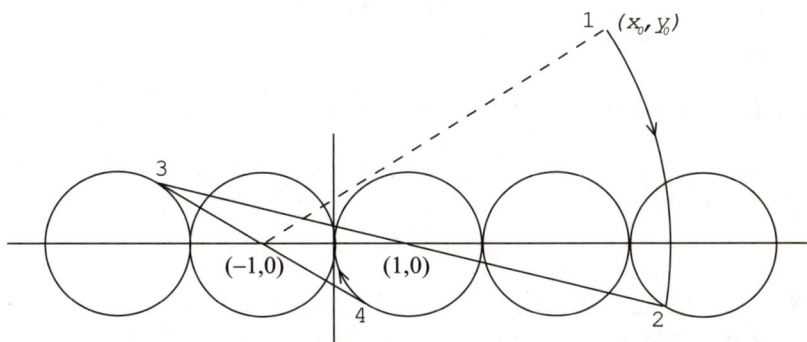

Figure 7.9 Once one of the unit circles is reached, the subsequent trajectory proceeds in a series of full-period switches until the path can be completed with a motion through an acute angle.

The strategy is now clear. Beginning with an arbitrary state, for instance at a point (x_0, y_0) in the upper half-plane, proceed clockwise along the circle with center $(-1, 0)$ for a time $t = \psi$ until intersecting the lower portion of one of the unit circles in the right half-plane shown in Figure 7.9, then proceed as shown, switching every π seconds until arriving at one of the two critical circles, where we can finish with a clockwise motion through an angle less than π. For (x_0, y_0) in the lower half-plane, proceed clockwise along the circle with center $(1, 0)$ for a time $t = \psi$ until intersecting the upper semicircle of one of the unit circles in the left half-plane. If doubts remain, experiment with graph paper and a compass (Exercise 7.37).

EXERCISES

7.1 Show that the work W of (7.1) done on the rolling cart by a successful nonnegative control signal $u \geq 0$ is at least 50.

Hint: $W = v(10)^2/2$ and (by the mean value theorem) $v(t) \geq 10$ at some intermediate t.

7.2 Show that by allowing negative forces (as in space docking), the cart of Figure 7.1 can be moved from $x = 0$ to $x = 100$ in $T = 10$ seconds with zero net work.

Hint: Use $u = a(5-t)$, or the pulse train control

$$u = \begin{cases} a & \text{if } 0 \leq t \leq T \\ 0 & \text{if } T < t < 10 - T \\ -a & \text{if } 10 - T \leq t \leq 10 \end{cases},$$

or the *bang-bang control* $u = \pm a$.

7.3 Show that the Euler-Lagrange equation applied to $N = v\dot{v} + \lambda v$ of (7.1)–(7.2) result in no information about the optimal control $u = \dot{v}$.

7.4 Of all nonnegative control signals u, show that impulsive control $u = 10\delta(t)$ yields the least work (7.1), namely, $W = 50$.

Outline: Consider a pulse control signal u that has value a for a short initial period $0 \leq t \leq T$ but 0 thereafter. Choose a so that the cart reaches $x = 100$ when $t = 10$. Show that in the limit, as $T \to 0$, $W \to 50$.

7.5 Show that the nonnegative linear control force (7.7) yields work $W \geq 112.5$.

Outline: Integrate (7.7) twice. Because $x(10) = 100$, $1 - u(0)/2 = 10\lambda/(12\epsilon)$. Because $u \geq 0$, $3 \geq u(0)$ But $v(10) = 30 - 5u(0)$.

7.6 Verify (7.10) and (7.11).

7.7 Show that the control force $u = a(10 - t)$ for the rolling cart of Figure 7.1 yields work $W = 112.5$. How much work is done when $u = a(10 - t)^n$?

7.8 Verify (7.19).

7.9* Using the cost

$$J = \int_0^{10} \epsilon^2 u^2 + u^2 v^2 \, dt$$

for the rolling cart of Figure 7.1, find a differentiable control u and the resulting work W done by u. Numerical methods may be necessary.

Exercises

7.10 Show that there is no upper bound on the work W that must be done in moving the cart of Figure 7.1 if we delay applying force until near the end of the allotted time.

Hint: Try the control force $u = a(t/10)^n$ or a delayed pulse.

7.11 Rework the rolling cart of §7.1 taking into account air resistance and viscous friction, that is, use the dynamic model

$$\dot{v} = -kv^2 - cv + u.$$

7.12 Show that by introducing the *states* $x_1 = x$, $x_2 = \dot{x}$, $x_3 = \ddot{x}$, the controlled system $x^{(3)} - 2x^{(2)} + 5t\dot{x} + 2x = u$ can be put into *state space* (vector) form $\dot{x} = F(t, x, u)$.

7.13 Verify (7.6).

7.14 Show that (7.15) cannot be optimal for the cost (7.8).

7.15 Show that the Lagrange equations for the augmented Lagrangian N of (7.23) lead to a contradiction. Why?

7.16 Establish (7.26) and (7.27).

7.17 Establish (7.30).

7.18 Complete the solution of OCP 3 in §7.4. Impose the constraints $v(0) = 0$, $E = 1$, and $v(1)$ unspecified to find the values of a, b, c.

Answer: $a = 0$, $b = \sqrt{3}$, $c = -\sqrt{3}/2$.

7.19 Verify (7.31).

7.20 (Fleming and Rishel) We wish to soft land on the Moon. Our initial altitude and velocity are $y(0) = y_0$ and $v(0) = v_0$, respectively. We are to touch down in $t = T$ seconds with terminal altitude and velocity $y(T) = v(T) = 0$. The thrust u of our retrorocket is proportional to the rate of fuel burn (i.e., $u = -k\dot{m}$, where m is the remaining mass of the vehicle plus fuel). The dynamics are of course $(mv)^{\cdot} = -mg + u$. Heavily penalizing fuel consumption with the cost

$$J = \int_0^T \dot{m}^2 \, dt,$$

show that the optimal continuously differentiable control force u is linear (i.e., $u = \alpha t + \beta$). Find the relationships among $y(0), v(0), m(0), k, \alpha, \beta, g, T$, and the empty mass M of the vehicle that must obtain if a crash is to be avoided.

7.21 You are to dock with a space station initially 100 km distant and receding at 2 km per second relative velocity. The thrust u of your motors is constant and is proportional to the rate of fuel burnoff (i.e., $u = \mp k\dot{m} = \mp kg$). You wish to dock in 2000 seconds with 0 relative velocity. How long should you accelerate, coast, and decelerate?

Hint: $\dot{p} = (mv)^{\cdot} = u$.

7.22 For the rolling cart of §7.1, let us severely penalize the use of unnecessary force but relax the requirement that we exactly reach $x = 100$ at $t = 10$ by means of the cost

$$J = \int_0^{10} 4\epsilon^4(100-x)^2 + u^2 \, dt$$

subject to the constraints $\dot{v} = u$, $\dot{x} = v$, and end conditions $x(0) = v(0) = 0$, plus the natural conditions at $t = 10$.
(a) Show that $x^{(4)} = -4\epsilon^4 x + 400\epsilon^4$.
(b) By experimenting with the value of ϵ, find a compromise control u that minimizes work W and error from final position $|x(10) - 100|$. Demonstrate that the work W is *extremely* sensitive to the values of the chosen parameters.

Answer: $x = 100(1 - \cosh \epsilon t \cos \epsilon t)$.

7.23 Is the elementary proof using (7.31) (that the at most one-switch control strategy of Figure 7.4 is indeed time-optimal) complete or is there more to be done?

7.24 Verify the important formula (7.33) for the unique trajectory of time-invariant linear controlled systems.

Outline: Note that the derivative of the convergent series of $n \times n$ matrices

$$X(t) = I + \frac{At}{1!} + \frac{A^2 t^2}{2!} + \frac{A^3 t^3}{3!} + \cdots$$

Exercises

is $\dot{X} = AX = XA$. Denote $X(t) = \exp(At)$, which is quite reasonable notation since $X(s+t) = X(s)X(t)$ and $X(0) = I$. Thus we may multiply through by the integrating factor $X(-t)$ to solve $\dot{x} - Ax = Bu$. Alternatively, show that (7.33) is a solution, and then apply Picard's theorem (Appendix B).

7.25* The set U of admissible controls of §7.7 is weak*-closed.

Outline: Suppose the sequence u_k in U is weak*-convergent to v, i.e., $x_k \hookrightarrow v$. Let x_k be the corresponding solutions to (7.33) and y the solution arising from the control v. Then by definition of weak*-convergence, using the notation of Exercise 7.24,

$$x_k(t) = X(t)a + \int_0^t X(t-\tau)Bu_k(\tau)\, d\tau \hookrightarrow$$

$$X(t)a + \int_0^t X(t-\tau)Bv(\tau)\, d\tau = y(t),$$

and so, in particular, $y(T) = b$. Since $\mathcal{B}$ is metrizable, sequentially closed sets are closed.

7.26* Prove that every compact convex subset K (of a Hausdorff topological vector space V) possesses extreme points.

Definition. A *face* F is a convex subset of K such that if f in F is an interior point of a line segment lying in K, then both endpoints of the line segment lie in F. The point e of a convex set K is an *extreme point* of K if it is impossible to find two distinct points x and y of K so that e can be written as the linear convex combination

$$e = tx + (1-t)y,\ 0 < t < 1,$$

that is, $\{e\}$ is a singleton face.

Outline: Apply Zorn's lemma to obtain a minimal closed face F. If F is not a singleton we can find a linear functional η such that $\eta(F)$ contains two distinct real numbers. The image of F under η is compact and convex and hence contains two extreme points. But the inverse image of an extreme point is a compact face.

7.27 Show that if a bang-bang control of $\mathcal{B}$ can be written in the form $u(t) = [u_1(t) + u_2(t)]/2$, where u_1, u_2 are in $\mathcal{B}$, then $u_1 = u_2 = u$ that is, bang-bang controls are extreme points of any subset of $\mathcal{B}$ to which they belong. Prove conversely that the extreme points of $\mathcal{B}$ are bang-bang. (Do the scalar case $r = 1$ first.)

7.28* Show that each extreme point u of the admissible controls U of §7.7 is in fact a bang-bang control.

Outline: Find a subset $M \subset [0, T]$ of positive measure where the extreme control u satisfies $-1 + \epsilon < u(t) < 1 - \epsilon$. Find a nonzero control v in the kernel of the map

$$I(v) = \int_M X(-\tau) B v(\tau) \, d\tau.$$

(See Exercise 7.24.) Extend v to be 0 off M. Normalize so that $|v|_\infty = 1$. Note that $u \pm \epsilon v \in U$ and $u = [(u + \epsilon v) + (u - \epsilon v)]/2$.

7.29* **(Bauer's Maximum Principle)** Show that the maximum value taken on by a convex function on a convex compact K must occur at an extreme point of K.

Definition. A realvalued function $\xi(x)$ on a convex subset K of a topological vector space is said to be *convex* if

$$\xi(ta + (1-t)b) \leq t\xi(a) + (1-t)\xi(b)$$

for all points $a, b \in K$ and $0 \leq t \leq 1$.

Outline: Let $\xi(K) = [a, b]$. Since $\xi(ta+(1-t)b) \leq \max(\xi(a), \xi(b))$, the maximum value b must occur at some boundary point k of K. We may assume $b = k = 0$. Separate the space with a continuous linear functional η so that K lies in the left half: $\eta(K) \leq 0$. Let N be the null space of η. Then $N \cap K$ is compact, convex, and contains $k = 0$ where the continuous convex function $\rho = \eta + \xi$ maximizes. If the underlying vector space V is finite dimensional, we may by induction on $\dim V$ we may assume ρ has a maximum value (viz. 0) at an extreme point e of $N \cap K$. But then $\xi(e) = \eta(e) = 0$. Thus e is also an extreme point of K, for if $e = (k_1 + k_2)/2$, $\eta(k_i) = 0$, thus k_i are in $N \cap K$.

Exercises

In the case that the underlying vector space V is infinite dimensional, order such separating η by inclusion applied to the intersections $N \cap K$. Choose a minimal η. Proceed as before to a contradiction.

7.30 Deduce the bang-bang principle (Theorem B) from Pontryagin's maximal principle (Theorem C).

Hint: Take $L = 1$.

7.31 Prove simple *transversality:* Suppose the vector variable $x = x(t)$ is a critical curve for the cost

$$J = \int_a^b L(t, x, \dot{x})\, dt$$

but the final value $x_k(b)$ is unspecified for some $1 \leq k \leq n$. Then

$$L_{\dot{x}_k}(b, x(b), \dot{x}(b)) = 0.$$

Second, by tracing through the proof of §5.4, show transversality also holds for augmented Lagrangians $N = L + \lambda M$ arising from an integral constraint.

7.32 Compute the total marketable production by the presumed optimal strategy of Example 1 in §7.8.

Answer: $Q = (q(0)/\alpha)e^{\alpha T - 1}$.

7.33 (Project) Taking into account reaction times and traffic patterns, how long should alternating green lights last to maximize traffic through an intersection?

7.34 Complete Example 2 of §7.8 by finding the appropriate T required and the resulting work W done. Assuming a is large enough so that the task is acheivable, show that the resulting control is optimal among all $0 \leq u \leq a$ for this fixed a.

7.35* (Controllability) Prove that if the $n \times nr$ controllability matrix

$$C = [B\ AB\ A^2B\ \cdots\ A^{n-1}B]$$

is of full rank n, then the time-invariant linear system $\dot{x} = Ax + Bu$ is *controllable*, that is, given any initial state x_0 and any target state x_1, there exists a piecewise continuous control $u = u^0$ that can steer the system from x_0 to x_1 in finite time. The converse also holds.

Outline: The *reachable* states of the form

$$x = \int_0^t e^{-A\tau} B u(\tau)\, d\tau$$

(for some t and piecewise continuous u) form a subspace R_t that increases with t, and so eventually, for some T, $R_t = R_T$ for all $t \geq T$. If not all states of $\mathbb{R}^n$ are reachable, then R_T must belong to some hyperplane $c^\top x = 0$ (i.e., $c^\top R_T = 0$). This means that

$$\int_0^t c^\top e^{A\tau} B u(\tau)\, d\tau = 0$$

for all t and piecewise continuous u, hence $c^\top e^{At} B u_0 = 0$ for all constant vectors u_0. But this is a Taylor series that vanishes on an open interval, hence $c^\top A^k B = 0$ for all k and hence $c^\top C = 0$. This argument is reversible via the Cayley-Hamilton theorem. Therefore, the column space of C equals R_T. Thus each target state is accessible in time T from any initial point if and only if C is of full rank n.

7.36 Show that Bushaw's system (Example 3) is controllable. Deduce that there is a unique path of least time from each point to the origin.

7.37 Using a compass and graph paper, construct the least time path from $(2, 3)$ to $(0, 0)$ in Bushaw's Example 3.

7.38 Rederive (7.7) via the maximum principle.

7.39 Rederive (7.19) via the maximum principle.

7.40 Verify (7.38).

7.41* Prove Corollary A to Theorem B in §7.6.

7.42 Deduce Corollary B to Theorem B in §7.6.

Exercises

7.43 Carefully establish that under the hypotheses of Theorem C, any solution $\lambda = \lambda(t)$ to the adjoint equation (7.35b) is piecewise continuously differentiable.

7.44 (**Optimality Principle**) Formulate hypotheses and prove that *every piece of an optimal trajectory is itself optimal*, otherwise one could construct a path of even lower cost (7.34b).

7.45 Rework Example 2 (OCP 1) of §7.8 for admissible controls $-a \leq u(t) \leq a$.

7.46 Carefully verify the calculations leading to (7.39).

7.47 Assuming the necessary differentiability, deduce the Euler-Lagrange equation (4.7) from the maximum principle when $L = L(y, y')$.

Hint: Let y' be the control u. Careful—see Exercise 7.54.

7.48 Deduce from the proof of the maximum principle that the optimal path x and adjoint variable λ satisfy the Hamilton-like equations
$$\dot{x} = H_\lambda \quad \text{and} \quad \dot{\lambda} = -H_x.$$

7.49 Deduce from the proof of the maximum principle that the adjoint variable λ is at each t either 0 or normal to contours of constant minimal cost $J(x(t)) = c_0$. Thus *the ith coordinate λ_i of the adjoint variable is minus the marginal rate of change of the cost J with respect to the ith coordinate x_i of x.*

7.50 Carefully adapt the proof of Theorem C to the variant of the maximum principle where the time T to target is specified but the final target value $x(T)$ is unspecified.

7.51 (**Transversality**) Expanding upon Exercise 7.50, because the target value $x(T)$ is not specified, we must have the *transversality condition* $\lambda(T) = 0$. Prove the more delicate version that if the i-th target coordinate b_i is unspecified, the corresponding coordinate of the adjoint variable $\lambda_i(T) = 0$. (Compare with Exercise 4.31.)

Outline: Let
$$\gamma(b; t) = \int_0^t L(x, u)\, ds,$$

where x is the optimal path from a to b in time T. If (say) the first coordinate b_1 of b is unspecified yet $\lambda_1(T) \neq 0$, then apply the inverse function theorem to the mapping $(b_1, t) \longmapsto \gamma(b_1, b_2^0, \ldots, b_n^0; t)$ at (b_1^0, T) to find a path of even lower cost in time T.

7.52 A third variant of the maximum principle is where the target b and time T to target is specified but where the initial point a is not. Prove that in this variant, $\lambda(0) = 0$.

7.53 For x a scalar variable, much of the proof of the maximal principle can be obviated. Show in one line that if $x = x^*$ and $u = u^*$ are optimal, $H(x^*, \lambda, u^*) \equiv 0$ implies $\dot{\lambda} = -H_x$.

7.54 Does (7.36) imply that $H_u = 0$?

Answer: No. Why not?

7.55 A drug is being ingested at the rate u so that the amount x in the gastrointestinal tract is governed by $\dot{x} = -ax + u$. The amount y in the bloodstream is given by $\dot{y} = ax - by$. What is the time-optimal strategy for reaching given bloodstream concentration without exceeding a given ingestion rate?

7.56 (Huygens' Isochrone) Find a planar curve where the time to slide down this curve from any point to the lowest point is identical.

Answer: The cycloid (4.25). See Figure 4.1 and [Kline]. A perfect timepiece!

Outline: The time needed to slide from $P = (x_0, y_0)$ to $Q = (\gamma^2 \pi/2, \gamma^2)$ is, via (4.22),

$$T = \frac{1}{\sqrt{2g}} \int_{y_0}^{\gamma^2} \frac{ds}{\sqrt{y - y_0}} = \frac{\gamma}{\sqrt{2g}} \int_{y_0}^{\gamma^2} \frac{dy}{\sqrt{\gamma^2 - y}\sqrt{y - y_0}}$$

$$= \frac{\gamma}{\sqrt{2g}} \int_0^{\beta} \frac{d\alpha}{\sqrt{\alpha}\sqrt{\beta - \alpha}} = \frac{\gamma}{\sqrt{2g}} \int_0^1 \frac{dz}{\sqrt{z}\sqrt{1 - z}} = \frac{\gamma \pi}{\sqrt{2g}}.$$

7.57 (Leibniz's Isochrone) Find the planar downward sloping curve such that a mass sliding on this curve has constant vertical velocity v.

Answer: In the coordinate system of Figure 4.1, $9v^2 x^2 = 8gy^3$.

Chapter 8
The LQ Problem

Let us now take on a more practical bent. How is optimal control actually used to control the familiar machinery around us? In many cases it is sufficient for good engineering results to employ linear approximate models with simple quadratic functions as performance indicies. This simple approach has been extremely successful in controlling many common devices. We will first introduce the approach, examine state feedback, define and characterize when linear systems are stable, state and prove the fundamental results about linear quadratic regulators, and finally work through a case study.

8.1 Problem Statement

Our task is to steer the time-invariant linear system

$$\dot{x} = Ax + Bu \qquad (8.1)$$

from a given *(initial)* state $x^0 = x(0)$ to the final *(target)* state $0 = x(T)$ in some finite time T. But even more, while doing so, our goal is to minimize the quadratic performance index

$$J = \int_0^T x^\top Q x + u^\top R u \, dt \; + \; x(T)^\top S x(T). \qquad (8.2)$$

Think of the task of moving a hard disk drive read-head from one track to another, quickly but without overshoot or excessive control effort.

Recall (Exercise 7.35) that if the system is controllable, that is, if

$$C = [B \; AB \; A^2 B \; \cdots \; A^{n-1} B] \qquad (8.3)$$

is of full rank n, then for every initial state x^0 and target state x^1, there are piecewise continuous[1] controls u steering the system (8.1) from x^0 to x^1 in finite time.

---
[1] A function is *piecewise continuous* if it possesses at most a finite number of discontinuities, all either removable or jump discontinuities.

For example, for our cart of Figure 7.1, since

$$\begin{vmatrix} \dot{x} \\ \dot{v} \end{vmatrix} = \begin{bmatrix} 0 & 1 \\ 0 & 0 \end{bmatrix} \begin{vmatrix} x \\ v \end{vmatrix} + \begin{vmatrix} 0 \\ -1 \end{vmatrix} u,$$

the controllability matrix

$$C = [B\ AB] = \begin{bmatrix} 0 & -1 \\ -1 & 0 \end{bmatrix}$$

is of full rank 2. Thus all steering tasks are possible by employing piecewise continuous controls.

Assumptions: Each of Q, R, S are symmetric and positive semidefinite.[2] The set of admissible controls U will consist of all bounded controls. Vector variables will be columns: x is a $n \times 1$ column and u is $r \times 1$. Thus A, Q, S are $n \times n$, B is $n \times r$, and R is $r \times r$.

The LQ Problem: Is there a control signal u that steers the linear (time-invariant) system (8.1) from the initial point $a = x(0)$ to the target 0 in a fixed, preassigned time $T > 0$ that minimizes the performance index (8.2)?

Result A. If a bounded control u^0 exists that steers the linear system (8.1) from x^0 to x^1 in time T, then there exists an optimal control u^* among the set of all such controls with $|u^*|_\infty \leq |u^0|_\infty$ that also minimizes the performance index

$$J = \int_0^T x^\top Q x\, dt. \tag{8.4}$$

Proof sketch.* The map

$$u \longmapsto J = \int_0^T x^\top Q x\, dt \tag{8.5}$$

is a continuous map on the weak*-compact subset of all controls u accomplishing the task with $|u|_\infty \leq |u^0|_\infty$ (Exercise 8.2). Thus the set of resulting costs J is compact and hence the minimal cost is achieved.

---

[2] A matrix X is *symmetric* when $X^\top = X$. A matrix X is *positive semidefinite* when $x^\top X x \geq 0$ for all vectors x.

8.1 Problem Statement

Result B. The optimal control guaranteed by Result A is bang-bang. It is in fact unique whenever the system (A, B) is *normal*, that is, whenever the matrix

$$N = [b_j \; Ab_j \; A^2 b_j \; \cdots \; A^{n-1} b_j] \tag{8.6}$$

is of full rank for all columns b_j of B.

Proof Sketch. The Hamiltonian for this problem is

$$H = -x^\top Q x + \lambda^\top (Ax + Bu).$$

With x and λ fixed, H maximizes exactly when $\lambda^\top B u$ maximizes, that is, when

$$h = \lambda^\top B u = \sum_{j=1}^{r} u_j \left(\sum_{i=1}^{n} \lambda_i b_{ij} \right)$$

maximizes. Therefore, by the maximum principle, since each $-a_j \leq u_j(t) \leq a_j$,

$$u_j(t) = a_j \operatorname{sgn} \left(\sum_{i=1}^{n} \lambda_i(t) b_{ij} \right) \tag{8.7}$$

whenever

$$\sum_{i=1}^{n} \lambda_i(t) b_{ij} \neq 0. \tag{8.8}$$

But normality guarantees that this summation cannot vanish on a set of t with positive measure (Exercise 8.17).

Example 1. (CVP 23) We are to steer the one-dimensional system $\dot{x} = -u$ from $x(0) = 1$ to 0 in finite time T, and at the same time minimize the performance index

$$J = \int_0^T x^2 \, dt. \tag{8.9}$$

The steering task is certainly doable since the system in controllable (even normal). More explicitly, take the (bang-bang) control $u = 1$, whereupon $x(t) = 1 - t$ and the target is reached when $T = 1$ with a cost of $J = 1/3$. By the preceding two results, there is a unique bang-bang optimal control u^* on $[0, 1]$ that accomplishes the task with least cost $|u^*| \leq 1$. In fact, $u = 1$ is this optimal control on $[0, 1]$ (Exercise 8.29).

However, there is no optimal control among the class U of all bounded controls since

$$u(t) = \begin{cases} -n & \text{if } 0 \leq t < 1/n \\ 0 & \text{otherwise} \end{cases} \tag{8.10}$$

will yield the performance $J = 1/(3n)$ yet $J = 0$ is not achievable (Exercise 8.3). So we must be careful with the import of Result A—it only guarantees an optimal control from among the *uniformly bounded* controls that achieve the task.

Questions. Do optimal controls exist that solve any LQ problem? Are they continuous or bang-bang? Can the task be completed in time T independent of the constraint matrices Q and R?

8.2 State Feedback

Let us visit the "Bell Labs" school of control theory. Rather than designing a control $u(t)$ ahead of time to accomplish a future task, why not "control on the fly" by using the present state $x(t)$ at time t to steer toward the target 0 in the time remaining? Put another way, design a clever matrix rule K that will read out the state $x(t)$, modify it in some way, then supply the result as the control signal u, that is, take

$$u(t) = K(t)x(t)$$

to form the *closed-loop* system

$$\dot{x} = Ax + BKx = (A + BK)x. \tag{8.11}$$

See Figure 8.1. This is called *state feedback control*.

At first hearing, feedback control sounds ideal since by closing the loop we can correct for in-course perturbations (e.g., unexpected wind gusts). In jargon *au courant*, "Closed-loop control is robust in the face of disturbances and model parameter uncertainties." The dream is to find a feedback strategy K so clever that we need not even understand the system to be controlled. But as can be seen from (8.11), once the trajectory is launched, it is completely determined by the initial state $x(0)$ and by the matrix $A + BK(t)$. We are shooting for the terminal state 0. There is nothing inherently better about feedback (online) versus optimal (offline) control—any

8.3 Stability

completed steering task $x = x(t)$ that was accomplished via state feedback rule $K(t)$ can be accomplished with the prescribed control signal $u(t) = K(t)x(t)$. In both cases, we must, before launch, specify either the control u or the feedback rule K.

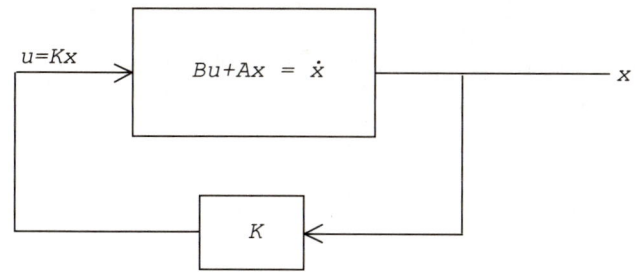

Figure 8.1 A closed-loop state feedback system.

Questions. Can state feedback solve the LQ problem? What LQ problems can be solved with *constant gain* feedback $K(t) = K$? For that matter, which target states are accessible using constant K?

8.3 Stability

A free-running system $\dot{x} = Ax$ is *(asymptotically) stable* if eventually every trajectory leads to 0, that is, for each initial point x^0,

$$x(t) = e^{At}x^0 \longrightarrow 0 \text{ as } t \longrightarrow \infty. \tag{8.12}$$

Result C. The linear system $\dot{x} = Ax$ is stable if and only if all eigenvalues of A lie in the open left half of the complex plane, Re $s < 0$.

Proof sketch 1. The real system $\dot{x} = Ax$ is stable if and only if it is a stable system in the complex vector space $\mathbf{C}^n$. We are then guaranteed a change of basis matrix P that brings A to its Jordan canonical form [Brown]. But

$$y(t) = Px(t) = Pe^{At}x^0 = Pe^{At}P^{-1}Px_0 = e^{PAP^{-1}}y^0, \tag{8.13}$$

and so it is sufficient to prove the result for a Jordan block $A = J$.

Since a Jordan block is the sum of a diagonal with a lower diagonal nilpotent matrix $J = \lambda I + N$,

$$y(t) = e^{\lambda I t} e^{Nt} y^0 = e^{\lambda t} e^{Nt} y^0, \tag{8.14}$$

where the series $\exp(Nt)$ breaks off at a polynomial.

Thus if $\operatorname{Re} \lambda < 0$, the exponential factor dominates giving $y(t) \longrightarrow 0$. Conversely, if all trajectories lead to 0, then in particular for the eigenvector v of J belonging to λ,

$$y(t) = e^{Jt} v = e^{\lambda t} v \longrightarrow 0,$$

and so $\operatorname{Re} \lambda < 0$.

Proof 2. In the special case where the eigenvectors v^i span, we may use them as a basis whereupon A diagonalizes and the solution to the system becomes

$$x(t) = e^{At} x^0 = x_1^0 e^{\lambda_1 t} v^1 + x_2^0 e^{\lambda_2 t} v^2 + \cdots + x_n^0 e^{\lambda_n t} v^n. \tag{8.15}$$

The result is now obvious.

Can feedback be used to stabilize an unstable system? There is this stunning result, a workhorse of control theory.

Theorem A. A controllable system can be stabilized by constant gain state feedback $u = Kx$. That is, given any linear time-invariant controllable system

$$\dot{x} = Ax + Bu, \tag{8.16}$$

there is a constant matrix K such that

$$\dot{x} = (A + BK)x \tag{8.17}$$

is stable. Even more, by a very clever choice of K, we can prescribe the eigenvalues of $A + BK$ as long as complex eigenvalues occur in conjugate pairs.

Proof.* We can quickly reduce to the case $r = 1$ of one scalar control as follows: Write the underlying vector space $V = \mathbf{R}^n$ as the

8.3 Stability

direct sum of indecomposable cyclic A-modules whereupon A takes the block-diagonal form

$$A = \begin{bmatrix} A_1 & 0 & \cdots & 0 \\ 0 & A_2 & 0 & \cdots \\ 0 & 0 & \ddots & \\ 0 & \cdots & & A_p \end{bmatrix} \qquad (8.18a)$$

with the corresponding block structure for B:

$$B = \begin{bmatrix} B_1 \\ \vdots \\ B_p \end{bmatrix}. \qquad (8.18b)$$

Note that each pair (A_i, B_i) is controllable. We have thereby reduced to the case where A is cyclic and indecomposable.

In this special case, the invariant subspaces of A are totally ordered. Thus one of the columns of B is cyclic for A and the remaining corresponding controls may be discarded as unnecessary for controlling the system.

The minimum number of control variables sufficient to control this system is the number of cyclic, indecomposable blocks of A.

Thus we have reduced to the case of one scalar control variable, that is,

$$B = \begin{vmatrix} b_1 \\ \vdots \\ b_n \end{vmatrix} = b. \qquad (8.19)$$

Since A is also left cyclic, there is a vector c so that

$$c^\top A^{n-1}, c^\top A^{n-2}, \ldots, c^\top A, c^\top$$

forms a basis. Thus there is a change of (row) basis such that

$$PAP^{-1} = \begin{bmatrix} -a_{n-1} & a_{n-2} & \cdots & -a_0 \\ 1 & 0 & \cdots & \\ 0 & \ddots & & \\ 0 & \cdots & 1 & 0 \end{bmatrix}, \qquad (8.20)$$

where the a_i are the coefficients of the characteristic polynomial

$$\phi(\lambda) = \lambda^n + a_{n-1}\lambda^{n-1} + \cdots + a_0. \qquad (8.21)$$

We may henceforth assume A is in this form (8.20). But by Exercise 8.9, there exists another change of basis P that fixes A yet maps b to $(1, 0, \ldots, 0)^\top$. Thus we have reduced to *control canonical form*:

$$\dot{x} = \begin{bmatrix} -a_{n-1} & a_{n-2} & \cdots & -a_0 \\ 1 & 0 & \cdots & 0 \\ 0 & \ddots & & \vdots \\ 0 & \cdots & 1 & 0 \end{bmatrix} x + \begin{bmatrix} 1 \\ 0 \\ \vdots \\ 0 \end{bmatrix} u. \quad (8.22)$$

So by taking state feedback

$$u = Kx = (k_1, k_2, \ldots, k_n)x, \quad (8.23)$$

we obtain the controlled system $\dot{x} = (A + BK)x$ with matrix

$$A + BK = \begin{bmatrix} -a_{n-1} + k_1 & a_{n-2} + k_2 & \cdots & -a_0 + k_n \\ 1 & 0 & \cdots & \\ 0 & \ddots & & \\ 0 & \cdots & 1 & 0 \end{bmatrix}. \quad (8.24)$$

Thus we may set the coefficients of the characteristic polynomial of $A + BK$ to any real values whatsoever.

Note that once the unnecessary controls are discarded, each distinct state feedback matrix K yields a distinct characteristic polynomial $\phi(\lambda)$, and so K is determined uniquely by the closed-loop eigenvalues.

8.4 The LQR Problem

A controllable, time-invariant linear, quadratically regulated system can steer any point x^0 to 0 in finite time. Such regulators can be applied widely, as we shall see in the following section. But first let us prove several basic results.

Theorem B. A time-invariant linear system

$$\dot{x} = Ax + Bu \quad (8.25\text{a})$$

possesses for almost all T and for each initial state x^0 an analytic[3] optimal control signal $u = u(t)$ that among all possible piecewise

---
[3] The entries can be expanded in Taylor series.

8.4 The LQR Problem

continuous controls minimizes the performance index

$$J = \int_0^T x^\top Q x + u^\top R u \, dt \tag{8.25b}$$

provided R is invertible. Moreover, u can be given as state feedback $u(t) = K(t)x(t)$ for a meromorphic matrix K.

Proof. Fix a time $T > 0$. Consider then the cost functional

$$I = \int_0^T L(x,u) - \dot{\lambda}^\top x - \lambda^\top [Ax + Bu] \, dt, \tag{8.26}$$

where

$$L(x,u) = x^\top Q x + u^\top R u. \tag{8.27}$$

Fix piecewise continuously differentiable curves $x(t), u(t), \lambda(t)$ with $x(0) = x^0$.

Introduce piecewise continuous perturbations $\eta = \eta(t)$ and $\zeta = \zeta(t)$. Rescale ζ until

$$\int_0^T \zeta^\top \zeta \, dt = 1. \tag{8.28}$$

Form the perturbed cost

$$I(\epsilon) = \int_0^T L(x+\epsilon\eta, u+\epsilon\zeta) - \dot{\lambda}^\top (x+\epsilon\eta) - \lambda^\top [A(x+\epsilon\eta) + B(u+\epsilon\zeta)] \, dt \tag{8.29}$$

and expand $I(\epsilon)$ in a two-term Taylor series with remainder:

$$I(\epsilon) = I(0) + I'(0)\epsilon + \frac{I''(c)\epsilon^2}{2!}. \tag{8.30}$$

But (Exercise 8.10)

$$I''(c) = \int_0^T \eta^\top Q \eta + \zeta^\top R \zeta \, dt \geq q_1 \int_0^T \eta^\top \eta + r_1 \int_0^T \zeta^\top \zeta \, dt \geq r_1 > 0, \tag{8.31}$$

where q_1 is the least (nonnegative) eigenvalue of Q and r_1 the least (positive) eigenvalue of R.

Thus if we can find x, u, λ making $I'(0) = 0$ via the Euler-Lagrange equation, then by (8.30) and (8.31) we have a global minimum of the augmented cost functional I. *In this special case, the*

Euler-Lagrange equations are both necessary and sufficient to guarantee a global minimum for the functional I.

The Euler-Lagrange equations for the integrand of (8.26) are

$$L_u - \lambda^\top B = 0, \qquad (8.32a)$$

and

$$L_x - \lambda^\top A = \dot{\lambda}^\top. \qquad (8.32b)$$

The first equation (8.32a) becomes (Exercise 8.13)

$$u = R^{-1} B^\top \lambda. \qquad (8.33a)$$

The second (8.32b) becomes (Exercise 8.14)

$$\dot{\lambda} + A^\top \lambda = Qx. \qquad (8.33b)$$

Note that if our solution $x = x(t)$ satisfies the right end condition $\lambda(T) = 0$, then by parts,

$$I = \int_0^T L(x, u) + \lambda^\top [\dot{x} - Ax - Bu]\, dt = J + \int_0^T \lambda^\top [\dot{x} - Ax - Bu]\, dt. \qquad (8.34)$$

Thus if, in addition to (8.33a) and (8.33b), our triple x, u, λ satisfies

$$\dot{x} = Ax + Bu, \qquad (8.35)$$

with $x(0) = x^0$ and $\lambda(T) = 0$, then this global minimum of I becomes the global minimum for the quadratic performance index J of (8.25b) and the pair (x, u) solves the LQR problem (8.25). Let us finish by showing that such a triple (x, u, λ) exists.

Our coupled system (8.33) and (8.35) becomes

$$\begin{vmatrix} \dot{x} \\ \dot{\lambda} \end{vmatrix} = \begin{bmatrix} A & BR^{-1}B^\top \\ Q & -A^\top \end{bmatrix} \begin{vmatrix} x \\ \lambda \end{vmatrix}, \qquad (8.36)$$

with solution given by the variation of paramenter formula (7.33) modified so that the trajectory is determined by the final states $x_T = x(T), \lambda_T = \lambda(T)$ at $t = T$:

$$\begin{vmatrix} x \\ \lambda \end{vmatrix} = e^{\begin{bmatrix} A & BR^{-1}B^\top \\ Q & -A^\top \end{bmatrix}(t-T)} \begin{vmatrix} x_T \\ \lambda_T \end{vmatrix} = \begin{bmatrix} \Gamma & \Delta \\ Z & H \end{bmatrix} \begin{vmatrix} x_T \\ \lambda_T \end{vmatrix}. \qquad (8.37)$$

8.4 The LQR Problem

Note that for any choice of terminal states, both x and λ (hence u) are continuously differentiable and in fact analytic.

The determinant of $\Gamma(t)$ is an analytic function of t that cannot vanish identically (since $\Gamma(T) = I$), and hence its zeros are isolated. By translation we may avoid T for which $\Gamma(0)$ fails to be invertible.

Let us choose $\lambda(T) = 0$, which is justified by transversality (Exercise 7.51) since we have not specified the target value $x(T)$. Then from (8.37) we have

$$x(t) = \Gamma(t)x_T, \tag{8.38a}$$
$$\lambda(t) = Z(t)x_T. \tag{8.38b}$$

Because $\Gamma(t)$ is invertible at $t = 0$, we may choose the value of x_T to achieve the initial state:

$$x(0) = \Gamma(0)x_T = x^0. \tag{8.39}$$

Moreover, combining (8.38a) and (8.37b), we have, except at possibly a finite number of points where $\Gamma(t)$ fails to be invertible,

$$u(t) = BR^{-1}B^\top \lambda(t) = BR^{-1}B^\top Z(t)\Gamma^{-1}(t)x(t), \tag{8.40}$$

and so the optimal control is given by state feedback.

Remark A. [Kwakernaak and Sivan] assert that Kalman proved that $\Gamma(t)$ is everywhere invertible. Examples will certainly bear that out (Exercise 8.19).

Remark B. The preceding strategy has not necessarily achieved a small target value $x(T)$. A slight modification of the proof will penalize large target values $x(T)$ (Exercise 8.20). Alternatively, we can take a fork in the proof at (8.38) where instead we apply the usual variation of parameter formula where the trajectory is determined by its *initial* states:

$$\begin{vmatrix} x \\ \lambda \end{vmatrix} = e^{\begin{bmatrix} A & BR^{-1}B^\top \\ Q & -A^\top \end{bmatrix} t} \begin{vmatrix} x^0 \\ \lambda_0 \end{vmatrix} = \begin{bmatrix} \Gamma & \Delta \\ Z & H \end{bmatrix} \begin{vmatrix} x^0 \\ \lambda_0 \end{vmatrix}. \tag{8.41}$$

It remains to show that given x^0 and for an appropriate initial choice of $\lambda_0 = \lambda(0)$, we can achieve the target $x(T) = 0$, that is,

$$\Gamma(T)x^0 + \Delta(T)\lambda_0 = 0. \tag{8.42}$$

Because $\det \Delta(t)$ is analytic in t, there are three cases:

(i) Either $\det \Delta(T) \neq 0$, in which case (8.42) holds and the LQR problem is solvable with target $x(T) = 0$, or

(ii) $t = T$ is an isolated zero of $\det \Delta(t)$, in which case the LQR problem is solvable for all final times near T, or

(iii) $\det \Delta(t) = 0$ for all t.

It is unkown to us if outcome (iii) may obtain.

Note that we must surrender any hope of state feedback when we insist on target $x(T) = 0$ (Exercise 8.15).

Remark C. Among other deductions, this remarkable Theorem B guarantees optimal controls that are analytic for almost any duration $T > 0$. In particular, a controllable system can be steered from any initial point to any target point in time T for almost all T arbitrarily small. See Exercise 8.16.

Sketch of the traditional proof of Theorem B. Assume for the moment that we have in hand a solution K to the *matrix Riccati equation*

$$\dot{K} = -A^\top K - KA - KBR^{-1}B^\top K + Q, \qquad (8.43)$$

where the $n \times n$ symmetric matrix $K = K(t)$ is subject to the terminal condition $K(T) = 0$. The existence of such a K can be deduced from the Hamilton-Jacobi-Bellman equation [Knowles, p. 90]. Set

$$u(t) = R^{-1}B^\top K(t)x(t). \qquad (8.44)$$

Let Φ be the transition matrix of the closed-loop system

$$\dot{x} = Ax + Bu = (A + BR^{-1}B^\top K)x, \qquad (8.45)$$

that is, if $x = \Phi x^0$, then

$$\dot{\Phi} = (A + BR^{-1}B^\top)\Phi. \qquad (8.46)$$

Therefore the cost becomes

$$J = \int_0^T x^\top Q x + u^\top R u \, dt = \int_0^T x^\top [Q + KBR^{-1}B^\top K]x \, dt$$

8.4 The LQR Problem

$$= \int_0^T x^\top[(A^\top + KBR^{-1}B^\top)K + \dot{K} + K(A + BR^{-1}B^\top)K]x \, dt$$

$$= \int_0^T x^{0\top}\Phi^\top[(A^\top + KBR^{-1}B)K + \dot{K} + K(A + BR^{-1}B^\top K)]\Phi x^0 \, dt$$

$$= -x^{0\top} \int_0^T \frac{d}{dt}[\Phi^\top K\Phi] \, dt \, x^0 = -x^{0\top} K(0)x^0. \tag{8.47}$$

We then choose $K(0)$ to be positive definite with the maximum possible minimum eigenvalue.

Surprisingly, the following "infinite horizon" version of Theorem B is more useful.

Theorem C. Assume the system (8.1) is controllable and that R is invertible. Then there is a unique constant matrix K so that the trajectory $x = x(t)$ of (8.1) induced by the state feedback control $u = R^{-1}B^\top K$ minimizes the performance index

$$J = \int_0^\infty x^\top Q x + u^\top R u \, dt. \tag{8.48}$$

Proof sketch. There is a unique positive definite symmetric matrix K satisfying the *time-invariant Riccati equation* [Knowles]

$$-A^\top K - AK + KBR^{-1}B^\top K = Q. \tag{8.49}$$

Then, aping the preceding proof (8.44)–(8.45), we see

$$\int_0^T x^\top Q x + u^\top R u \, dt = \cdots = x^{0\top} \left(\Phi^\top(t) K \Phi(t) \Big|_0^T \right) x^0.$$

But the matrix K stablizes the system so that $\Phi(t) \longrightarrow 0$ as $t \longrightarrow \infty$. Thus

$$J = \int_0^\infty x^\top Q x + u^\top R u \, dt = -x^{0\top} K x_0.$$

Because any optimal control must also be optimal when starting further along in the trajectory $x^1 = x(t_1)$, detailed analysis shows K yields the least cost for any initial starting state x^0.

8.5 A Tracking Servo

Consider a *servo mechanism,* a device that multiplies the force applied by its operator—for example, the hydraulic control levers of a backhoe, the waldo manipulars used to handle radioactive materials, or the yoke/control surfaces of aircraft. In all these cases, with proper translation and rescaling, the output movement tracks the operator input movement, more or less. That is the rub—more or less. The mechanism itself has its own dynamics induced by (say) the inertia of the moving parts and the inherent dynamics of the sensors and actuators.

We assume that the internal response state x of the servo to a control input u is to good approximation given by a linear model

$$\dot{x} = Ax + Bu. \qquad (8.50)$$

Goal: To improve the accuracy with which the measurement $y = Cx$ tracks the control input $u = r$.

Example 3. Consider the one degree-of-freedom case

$$m\ddot{x} + c\dot{x} + kx = u, \qquad (8.51)$$

where say x is the extension of a hydraulic cylinder compressing a spring in response to a movement u of the control lever. To improve comprehension let us set mass, damping, and stiffness $m = c = k = 1$. Our goal is to provide feedback to improve the performance index

$$J = \int_0^\infty e^2 \, dt = \int_0^\infty (x - u)^2 \, dt \qquad (8.52)$$

in response to the *step input* $u = 1$ given that the mechanism starts from rest: $x(0) = 0$ and $\dot{x}(0) = 0$.

By going over to the error variable $e = x - u = x - 1$, the model (8.51) becomes (Exercise 8.21)

$$\ddot{e} + \dot{e} + e = 0 \qquad (8.53)$$

with solution

$$e = -e^{-t/2}[\cos \omega t + (1/2\omega) \sin \omega t], \quad \omega = \sqrt{3}/2, \qquad (8.54)$$

8.5 A Tracking Servo

and performance
$$J = \int_0^\infty e^2 \, dt = 1. \tag{8.55}$$

Can we improve on this performance?

Solution Attempt 1. Let us try state feedback $u = 1 - \kappa e$ as in Figure 8.2.

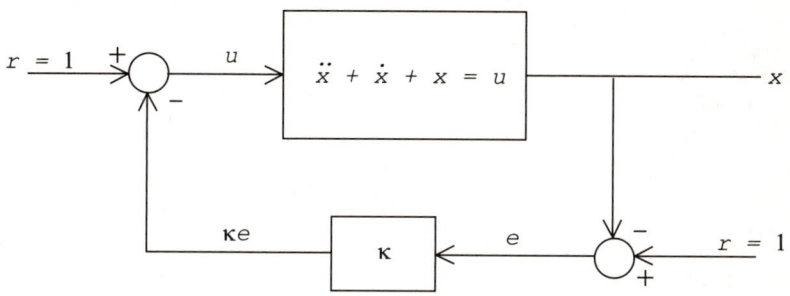

Figure 8.2 An attempt to improve the tracking of a servo with proportional state feedback.

The closed-loop system is
$$\ddot{e} + \dot{e} + (1 + \kappa)e = 0 \tag{8.56}$$

with solution
$$e = -e^{-t/2}[\cos \omega t + (1/2\omega) \sin \omega t], \quad \omega = \sqrt{3 + 4\kappa}/2, \tag{8.57}$$

and performance
$$J = \int_0^\infty e^2 \, dt = \frac{1}{2} \frac{\kappa + 2}{\kappa + 1} \tag{8.58}$$

(Exercise 8.22). Thus the performance can be improved by taking larger and larger gain κ at the price of more and more violent feedback signal and rapid oscillations. See Figure 8.3. There is no optimal such gain κ.

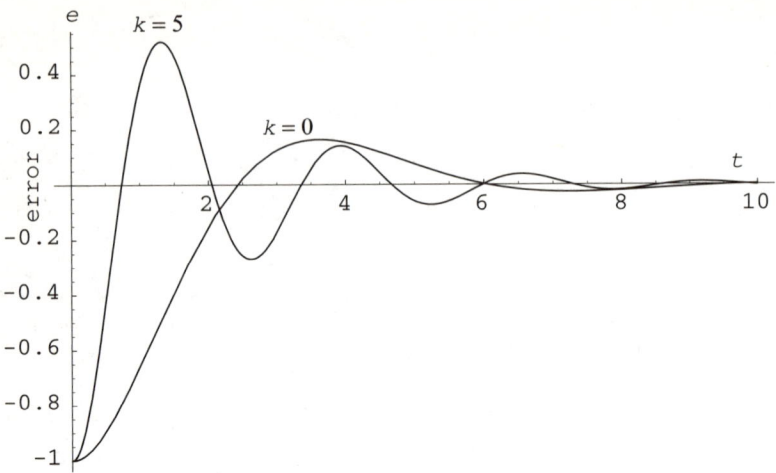

Figure 8.3 Although the performance index J is lowered, the motion is more violent and oscillatory with increasing proportional gain κ.

Solution Attempt 2. For hydraulics it is quite simple to control damping with little control effort, to control c where $\ddot{x} + c\dot{x} + x = u$. In this case, error $e = x - 1$ satisfies

$$\ddot{e} + c\dot{e} + e = 0, \quad 0 < c \leq 2, \tag{8.59}$$

with solution

$$e = -e^{-ct/2}[\cos \omega t + (c/2\omega)\sin \omega t], \quad \omega = \sqrt{4 - c^2}, \tag{8.60}$$

and performance index (Exercise 8.23)

$$J = \int_0^\infty e^2 \, dt = \frac{1}{2}(1 + \frac{1}{c}). \tag{8.61}$$

Thus the performance index minimizes when the system is critically damped, when $c = 2$, giving $J = 3/4$. Critical damping results in no *overshoot*. See Figure 8.4. However, the slow response (long *rise time*) may be unacceptable. What would happen if we combined both Attempts 1 and 2?

8.5 A Tracking Servo

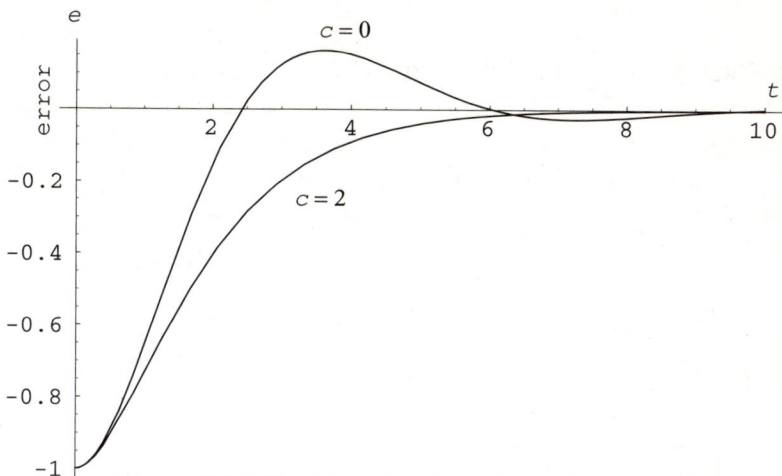

Figure 8.4 Critical damping yields the performance index $J = 3/4$ with no overshoot, but the response is sluggish.

Solution Attempt 3. Let us feedback $u = 1 - ke - (c-1)v$ where v is the velocity $v = \dot{x}$. The step tracking error $e = x - 1$ becomes
$$\ddot{e} + c\dot{e} + (k+1)e = 0. \tag{8.62}$$
In state space this is
$$\begin{vmatrix} \dot{e} \\ \dot{v} \end{vmatrix} = \begin{bmatrix} 0 & 1 \\ -1 & -1 \end{bmatrix} \begin{vmatrix} e \\ v \end{vmatrix} + \begin{vmatrix} 0 \\ 1 \end{vmatrix} u, \tag{8.63a}$$
$$u = -(c-1)v - ke. \tag{8.63b}$$

See Figure 8.5.

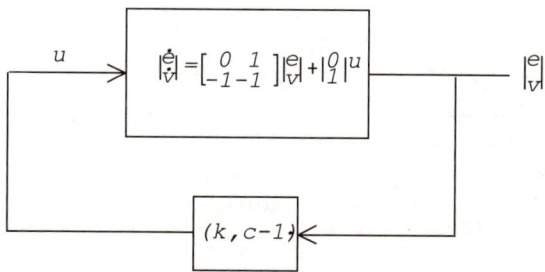

Figure 8.5 Closed-loop state feedback.

According to Theorem C, there is an optimal state feedback that will minimize

$$J = \int_0^\infty e^2 + \epsilon u^2 \, dt. \tag{8.64}$$

Is (8.63b) this optimal state feedback?

A choice of $k = 5$ and $c = 4.6$ yields the excellent apparent performance shown in Figure 8.6 (Exercise 8.24). There must be a price. What is it? The answer lies in the excessive force needed to accomplish this tracking (see Exercise 8.25).

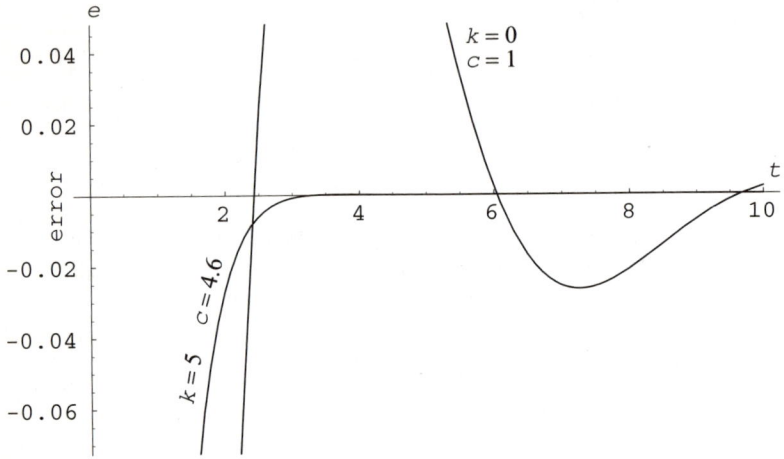

Figure 8.6 At first glance, performance appears vastly improved. There must be a hidden price.

There is extensive engineering literature on the LQR problem. MATLAB automates the search for the feedback matrix with the command LQR (Exercise 8.27).

EXERCISES

8.1 Is the system

$$\dot{x} = \begin{bmatrix} 0 & 1 & 0 \\ 0 & 0 & 1 \\ 0 & -3 & -2 \end{bmatrix} x + \begin{bmatrix} 0 \\ 0 \\ 1 \end{bmatrix} u$$

controllable?

Answer: No.

Exercises

8.2* Prove that the map (8.5) is indeed sequentially weak*-star continuous.

Hint: Use ideas from Exercise 7.25.

8.3 Show that the control of (8.10) does indeed yield $J = 1/3n$. Show that this is optimal among all controls $|u| \le n$.

8.4 Show that the system of Examples 1 and 2 is controllable. Are all one-dimensional systems controllable?

8.5 Verify (8.10) and that u is indeed optimal among bounded piecewise continuous controls on $[0, 1]$.

Hint: Apply Theorem B.

8.6* Show by an example that the map

$$u \longmapsto \int_0^T u^\top R u \, dt$$

cannot be continuous from the weak*-topology on the closed unit ball of $L^\infty[0,T]$ to $\mathbf{R}$.

Suggestion: Use $u_k(t) = \sin nt$ for $T = \pi$, $T = \pi$, $R = 1$.

8.7 Given an example of a control and a time-invariant linear system where the states are not everywhere differentiable.

8.8 Show that every trajectory of a constant gain K state feedback controlled system (8.11) is an analytic (vector-valued) function of time t.

8.9 Given any $n \times n$ matrix A and any two cyclic vectors b, c for A, there exists an invertible change of basis matrix P such that

$$PAP^{-1} = A \quad \text{and} \quad Pb = c.$$

Suggestion: Let P change from basis $b, Ab, A^2b, \ldots, A^{n-1}b$ to the basis $c, Ac, A^2c, \ldots, A^{n-1}c$

8.10 Verify that the second derivative $I''(c)$ of (8.30) is indeed given by (8.31).

8.11 Show that a quadratic form $F(x) = x^\top Q x$ is a convex function when Q is symmetric and positive semidefinite.

Preferred outline: The bilinear nonnegative form $\langle x, y \rangle = y^\top Q x$ has $2\langle x, y \rangle \leq \langle x, x \rangle + \langle y, y \rangle$ and hence $F(x) = \langle x, x \rangle$ is convex.

8.12 Show that the cost functional I of (8.26) is convex.

Hint: Show that the integrand is convex by noting it is the sum of linear terms plus quadratic forms.

8.13 Verify (8.33a).

8.14 Verify (8.33b).

8.15 Show that in the proof of Theorem B, it is impossible that both $x(T) = 0$ and $\lambda(T) = 0$.

8.16 Prove that in the proof of Theorem B, $\det \Delta(T) \equiv 0$ implies there exists an analytic row vector $c = c(t)$ with $c\Delta(t) \equiv 0$ and $c(0) \neq 0$.

Outline:

8.17 Complete the proof of Result B by showing that normality of the system (A, B), a much stronger requirement than controllability, guarantees that the optimal (bang-bang) control is unique.

8.18 Redo Example 2 for any time period T. Show that the cost

$$\int_0^T u^2 \, dt \geq 1/\sqrt{T}.$$

8.19 Work out a scalar example ($n = 1$) of (8.37). Show that $\Gamma(t)$ is invertible for all t.

8.20 Redo the first proof of Theorem B after penalizing the final target size with the cost

$$J = \int_0^T x^\top Q x + u^\top R u \, dt + x(T)^\top S x(T),$$

where S is positive semidefinite.

Exercises

8.21 Verify (8.53)–(8.55).

8.22 Verify (8.56)–(8.58).

8.23 Verify (8.59)–(8.61).

8.24 Obtain the graphs of Figure 8.6 via *Mathematica* or MATLAB.

8.25 Using the simulation developed in Exercise 8.24, graph the force levels $F = m\ddot{x} = \ddot{x}$ needed to achieve the errors shown in Figures 8.3, 8.4, and 8.6.

8.26 Find a state feedback that stabilizes $\ddot{x} + \dot{x} - 2x = u$.

8.27 Using MATLAB's LQW, find the optimal Liapunov matrix for the problem (8.63). Graph the resulting error response.

8.28* Is the PDE system $u_t = u_{xx}$ subject to $u(0) = 0 = u(1)$ stable?

8.29 Verify that $u \equiv 1$ is indeed the optimal control on $[0,1]$ with $|u| \leq 1$ in Example 1 of §8.1.

8.30 Prove that an asymptotically stable finite dimensional matrix system $\dot{x} = Ax$ is in fact *exponentially stable*, that is, there are positive constants a and c so that

$$|x(t)| = |e^{At}x^0| \leq ce^{-at}|x^0|.$$

8.31* Display an asymptotically stable linear system $\dot{x} = Ax$ that is not exponentially stable.

Hint: Consider the diagonal matrix $A = \text{diag}[1/n]$ on ℓ^2.

8.32 Find the optimal control $|u| \leq 1$ that steers the one-dimensional system $\dot{x} = -u$ from $x(0) = 1$ to 0 in some finite time T and minimizes the quadratic performance index

$$J = \int_0^1 x^2 + u^2 \, dt.$$

Of course we can achieve the task with the constant control signal $u = 1$ with cost $J = 4/3$ in time $T = 1$. But is this optimal? Is there some other control that yields a lower cost if we shorten or lengthen the time period T?

8.33 Formulate conditions under which the Lagrange equations must hold at a local minimum $q = q(t)$ of an improper integral cost

$$J = \int_0^\infty L(t, q, \dot q)\, dt.$$

8.34* (Project) [J. A. Mirrlees] Earning ability α is distributed throughout the U. S. population according to some (say) log-normal-like probability distribution $f(\alpha)$, so that an individual of ability α that works p percent of a day will earn $p\alpha$ dollars for that day. This person's *utility* (well-being) u is a function of take-home daily pay y less some function of the time spent working, for instance, $u(y, p) = a \log y + b \log(1 - p)$.

The goal is to design a tax structure that redistributes wealth by specifying an after-tax income $y = c(p\alpha)$ that maximizes society's total utility

$$J = \int_0^\infty u(c(p\alpha), p) f(\alpha)\, d\alpha$$

while conserving total wealth

$$\int_0^\infty c(p\alpha) f(\alpha)\, d\alpha = \int_0^\infty p\alpha f(\alpha)\, d\alpha.$$

Once the the tax plan c is codified, individuals will of course maximize their own utility, giving that $p = p(\alpha)$.

Find this optimal tax structure.[4]

---

[4] Sorry, but the Nobel has already been awarded for this optimization problem.

Chapter 9
Weak Sufficiency

All of our optimization theory to this point consists of necessary conditions—the Euler-Lagrange equation and the maximum principle at best yield candidates for an extremal path or an optimal control. It is not hard to imagine catastrophic engineering failures that would result from choosing a path that maximized rather than minimized some important performance index; we need guarantees that certain extremals are in fact maxima or minima. In this chapter we will first distinguish between weak and strong extrema and then abstract the problem to functionals on normed spaces (where the variations that give name to this subject are finally defined). We specialize our abstractions to cost functionals and work through two revealing examples. We then obtain sufficiency for a very special case and deduce a sufficiency theorem for weak local minima from this special case. We follow the traditional approach laid out in the classic text by Gelfand and Fomin.

9.1 Weak versus Strong Extrema

Let us restate our basic question: Which path $y = y(x)$ chosen from a prescribed class C of curves[1] minimizes the cost functional[2]

$$J[y] = \int_a^b L(x, y(x), y'(x))\, dx? \qquad (9.1)$$

Suppose we have a candidate path

$$y = y^*(x) \qquad (9.2)$$

for minimal cost. If for all paths

$$y = y^*(x) + \eta(x) \qquad (9.3)$$

---

[1] In application, the class C often consists of all piecewise continuously differentiable or twice-differentiable paths with specified initial and terminal points.

[2] We assume that $L(x, y, z)$ has continuous first partial derivatives everywhere.

in C with
$$\|\eta\|_0 = \sup_{a \le x \le b} |\eta(x)| < \delta \tag{9.4}$$
we have
$$J[y^* + \eta] \ge J[y^*], \tag{9.5}$$
then we say our candidate $y = y^*(x)$ is a *strong local minimum* because we have admitted *all* nearby paths in C that differ little in norm. Sometimes both the cost functional J and class C of admissible paths y are convex, giving that the local minimum is global (Exercise 1.27).

A *weak* local minimum is where (9.5) need hold only for variations $y - y^*(x) = \eta(x)$ in C in which
$$\|(\eta, \eta')\|_0 = \|\eta\|_0 + \|\eta'\|_0 < \delta \tag{9.6}$$
for some small δ. We look only at variations that are small both in value and derivative.

These two notions of weak and strong local minima do not coincide. See Exercise 9.1 and Example 2.

Strong local minima are clearly weak local minima. As we saw in Chapter 4, the Euler-Lagrange equation will find candidate local extremals. But are these candidate extremals actual minima of either type? It is revealing to take a more abstract view.

9.2 First and Second Variations

Let V be a *normed (real) vector space*, that is, a vector space with a nonnegative realvalued function $\|\cdot\| : V \longrightarrow \mathbb{R}$ satisfying

(i) $\|v\| \ge 0$ for all $v \in V$, with equality only when $v = 0$

(ii) $\|cv\| = |c|\,\|v\|$ for all $c \in \mathbb{R}$ and $v \in V$

(iii) $\|u + v\| \le \|u\| + \|v\|$ for all $u, v \in V$.

Such norms often arise from an *inner product* on V; see Exercise 9.2.

Because in application V is a function space, it has been traditional to refer to realvalued functions on V as *functionals*.

A functional
$$A : V \longrightarrow \mathbb{R} \tag{9.7a}$$

9.2 First and Second Variations

satisfying
$$A[cv] = cA[v] \text{ for all } c \in \mathbb{R}, v \in V, \tag{9.7b}$$

$$A[u+v] = A[u] + A[v] \text{ for all } u, v \in V \tag{9.7c}$$

is called a *linear functional* on V.

A function
$$B : V \times V \longrightarrow \mathbb{R} \tag{9.8a}$$
satisfying
$$B[u,v] = B[v,u] \text{ for all } u, v \in V, \tag{9.8b}$$

$$B[cu,v] = cB[u,v] \text{ for all } c \in \mathbb{R} \text{ and } u, v \in V, \tag{9.8c}$$

$$B[u+v,w] = B[u,w] + B[v,w] \text{ for all } u, v, w \in V \tag{9.8d}$$

is called a *symmetric bilinear functional*.

A symmetric bilinear functional B is *positive semidefinite* when
$$B[v,v] \geq 0 \text{ for all } v \in V \tag{9.9}$$

and *positive definite* when equality holds in (9.9) only when $v = 0$. Most important for our intended application, the symmetric bilinear form B is *uniformly positive (coercive)* if for some positive constant γ,
$$B[v,v] \geq \gamma \|v\|^2 \quad \text{for all } v \in V. \tag{9.10}$$

For example, an inner product is coercive on its own induced norm with $\gamma = 1$; see Exercise 9.2.

Let H be a subspace of V and let $C = v_0 + H$ be the residue class (coset) of v_0 modulo H. Consider a functional J on V. We say J is H-*differentiable* at v_0 if for some linear functional A and functional E on H, we have for $h \in H$,
$$J[v_0 + h] = J[v_0] + A[h] + E[h]\|h\|, \tag{9.11a}$$
where
$$\lim_{\|h\| \to 0} E[h] = 0. \tag{9.11b}$$

This *first variation* A of J at v_0 is clearly unique (Exercise 9.3).

We say that the differentiable functional J is *twice H-differentiable* at v_0 if in addition there exists a symmetric bilinear form B and a functional E on H so that

$$J[v_0 + h] = J[v_0] + A[h] + B[h,h] + E[h]\|h\|^2, \qquad (9.12a)$$

where
$$\lim_{\|h\| \to 0} E[h] = 0. \qquad (9.12b)$$

The *quadratic functional* Q given by

$$Q[h] = B[h,h] \qquad (9.13)$$

is called the *second variation* of J at v_0. It is also clearly unique (Exercise 9.4).

We have the following familiar formulas for the first and second variations: Fix $h \in H$ and let

$$J(\epsilon) = J[v_0 + \epsilon h]. \qquad (9.14)$$

Then from (9.12),

$$J(\epsilon) = J(0) + \epsilon A[h] + \epsilon^2 Q[h] + \epsilon^2 E[\epsilon h]\|h\|^2, \qquad (9.15)$$

that is,
$$A[h] = J'(0), \qquad (9.16a)$$

$$Q[h] = J''(0)/2!. \qquad (9.16b)$$

Classical Notation. The first and second variations A and Q are traditionally denoted by

$$\delta J = A \text{ and } \delta^2 J = Q. \qquad (9.17)$$

But why are all these abstractions of interest?

Result A. (necessity) If J takes on a local minimum value on $C = v_0 + H$ at v_0, then its first variation at v_0 is the zero functional and its second variation is positive semidefinite.

9.3 In Application

Proof. If J has a local minimum at v_0, then from (9.11), for each fixed h and positive ϵ,

$$0 \leq \frac{J(\epsilon) - J(0)}{\epsilon} = A[h] + \|h\|E[\epsilon h],$$

where

$$\lim_{\epsilon \to 0} E[\epsilon h] = 0.$$

Thus $A[h] \geq 0$. Replacing h by $-h$ yields $A[h] \leq 0$. Thus $A[h] \equiv 0$. But then from (9.12),

$$0 \leq \frac{J(\epsilon) - J(0)}{\epsilon^2} = Q[h] + \|h\|^2 E_1[\epsilon h],$$

where $\lim_{\epsilon \to 0} E_1[\epsilon h] = 0$. Thus $Q[h] \geq 0$.

Result B. (sufficiency) If the first variation is zero and the second variation is uniformly positive definite at v_0, then J has a local (isolated) minimum on C at v_0.

Proof. Suppose $A = \delta J = 0$ and

$$Q[h] = B[h, h] \geq \gamma \|h\|^2 \text{ for all } h \in H \text{ and some } \gamma > 0.$$

Then from (9.12), for $\|h\| \neq 0$,

$$\frac{J[v_0 + h] - J[v_0]}{\|h\|^2} - E[h] = \frac{B[h, h]}{\|h\|^2} \geq \gamma.$$

Thus for all small $\|h\|$,

$$J[v_0 + h] - J[v_0] \geq \frac{\gamma}{2}\|h\|^2 \geq 0.$$

Let us now apply these abstractions to guarantee that certain extremals obtained via the Euler-Lagrange equation are indeed paths of minimum cost. We will accomplish this by showing the second variation of their cost to be uniformly positive definite with respect to the *Dirichlet* norm.

9.3 In Application

Let us specialize the abstractions of the preceding section to our cost J of (9.1). Assume the Lagrangian L of our original cost (9.1) is at least thrice continuously differentiable. We consider variations of the cost functional

$$J[v] = \int_a^b L(x, y(x), y'(x))\, dx$$

on the normed space V of all pairs

$$v = (y, y'),$$

where y is absolutely continuous with bounded derivative, under the norm of (9.4) and (9.6):

$$\|v\|_0 = \|y\|_0 + \|y'\|_0.$$

Our admissible variations h of the particular path

$$v_0 = (y^*, y^{*\prime}) \tag{9.18a}$$

will lie in some subspace $H \subset V$ of pairs

$$h = (\eta, \eta'). \tag{9.18b}$$

We now expand $J(t) = J[v_0 + th]$ in a three-term Taylor series in t about the center $t = 0$, then set $t = 1$ to obtain (Exercise 9.5)

$$J[v_0 + h] = J[v_0] + A[h] + Q[h] + E[h], \tag{9.19a}$$

where

$$J[v_0] = \int_a^b L\, dx, \tag{9.19b}$$

$$A[h] = \int_a^b L_y \eta + L_{y'} \eta'\, dx, \tag{9.19c}$$

$$Q[h] = \frac{1}{2!} \int_a^b L_{yy}\eta^2 + 2L_{yy'}\eta\eta' + L_{y'y'}\eta'^2\, dx, \tag{9.19d}$$

$$E[h] = \frac{1}{3!} \int_a^b L_{yyy}\eta^3 + 3L_{yyy'}\eta^2\eta' + 3L_{yy'y'}\eta\eta'^2 + L_{y'y'y'}\eta'^3\, dx, \tag{9.19e}$$

9.3 In Application

where the three integrands of (9.19b)–(9.19d) are evaluated along $(x, y^*(x), y^{*\prime}(x))$ and where the integrand of (9.19e) is evaluated along $(x, y^*(x) + t^*\eta(x), y^{*\prime}(x) + t^*\eta'(x))$ for some $0 < t^* < 1$.

Note that because the Lagrangian L is thrice continuously differentiable, when we restrict our variations to a neighborhood $\|h\|_0 \leq \epsilon_0$, L and all its partials in (9.19e) are uniformly bounded by (say) c, giving that (Exercise 9.6)

$$|E[h]| \leq c(b-a)\|h\|_0^3, \qquad (9.20)$$

guaranteeing that A and Q in (9.19a) are the first and second H-variations of J, respectively, at v_0. We will need, however, a more delicate estimate.

Lemma A. Let c be a bound for all the third partial derivatives of L on the set of all points of the form $(x, y^*(x) + \eta(x), y^{*\prime}(x) + \eta'(x))$, where $a \leq x \leq b$ and $\|h\|_0 = \|\eta\|_0 + \|\eta'\|_0 \leq \epsilon_0$. Assume $\eta(a) = 0$. Then for the functional E of (9.19e),

$$|E[h]| \leq m\|h\|_0 \cdot \|\eta'\|_2^2, \qquad (9.20a)$$

where

$$m = c\frac{(b-a)^2 + 2\sqrt{2}(b-a) + 2}{12} \qquad (9.20b)$$

and

$$\|\eta'\|_2 = \sqrt{\langle \eta', \eta' \rangle} = \left(\int_a^b \eta'^2 \, dx\right)^{1/2}. \qquad (9.20c)$$

Proof. By the Cauchy-Schwarz inequality (Exercise 9.2),

$$3!|E[h]| \leq \int_a^b c|\eta|^3 + 3c|\eta|^2|\eta'| + 3c|\eta||\eta'|^2 + c|\eta'|^3 \, dx$$

$$= c\int_a^b (|\eta| + |\eta'|)^3 \, dx \leq c\|h\|_0 \int_a^b (|\eta| + |\eta'|)^2 \, dx$$

$$= c\|h\|_0 \int_a^b \eta^2 + 2|\eta||\eta'| + \eta'^2 \, dx$$

$$\leq c\|h\|_0 \left[\int_a^b \eta^2 \, dx + 2\left(\int_a^b \eta^2 \, dx \int_a^b \eta'^2 \, dv\right)^{1/2} + \int_a^b \eta'^2 \, dx\right]. \qquad (9.21)$$

Chapter 9. Weak Sufficiency

The lemma now follows easily from the very useful inequality (Exercise 9.17)

$$\int_a^b \eta^2 \, dx \le \frac{(b-a)^2}{2} \int_a^b \eta'^2 \, dx. \tag{9.22}$$

Remark. The second variation Q of (9.19d) is an integral of the quadratic form

$$q(\eta, \eta') = (\eta, \eta') \begin{bmatrix} L_{yy} & L_{yy'} \\ L_{y'y} & L_{y'y'} \end{bmatrix} \begin{vmatrix} \eta \\ \eta' \end{vmatrix}. \tag{9.23}$$

Occasionally this quadratic integrand q is of one signature (see Exercises 9.7–9.15) everywhere along the extremal, or, even more rarely, of one signature independent of the path, as in the following example.

Example 1. (CVP 1, §4.1) The shortest continuously differentiable path $y = y(x)$ connecting $(0,0)$ to $(b,0)$ is an extremal of the arclength

$$J = \int_0^b \sqrt{1 + y'^2} \, dx. \tag{9.24}$$

For this Lagrangian $L = \sqrt{1 + y'^2}$ we have $L_{yy} = 0$, $L_{yy'} = L_{y'y} = 0$, and

$$L_{y'y'} = 1/(1 + y'^2)^{3/2}.$$

Thus when H is the space of all pairs $h = (\eta, \eta')$ with η continuously differentiable and $\eta(0) = 0 = \eta(b)$,

$$\delta^2 J[h] = \frac{1}{2!} \int_0^b \frac{\eta'^2 \, dx}{(1 + y'^2)^{3/2}} \ge 0. \tag{9.25}$$

Thus the second H-variation is positive semidefinite, in fact positive definite, for if $\delta^2 J[h] = 0$, then $\eta'(x) \equiv 0$, hence $\eta = 0$ since $\eta(0) = 0$. This is not enough to assure that an extremal y^* is a weak local minimum of the cost J; it guarantees only that J is locally minimized by y^* along each 'ray' $y = y^* + \epsilon\eta$, $|\epsilon| < \epsilon_0(\eta)$, for each fixed η, a difficulty unappreciated by some authors.

9.3 In Application

Recall that $y^* = 0$ is an extremal of (9.24). For this extremal the value of the second variation (9.25) becomes

$$\delta^2 J[h] = \frac{1}{2!} \int_0^b \eta'^2 \, dx = \frac{\|\eta'\|_2^2}{2}. \tag{9.26}$$

But then from the estimate (9.20a), for all $\|h\|_0 < m/4$, we see

$$J[0+h] - J[0] = \delta^2 J[h] + E \geq \frac{\|\eta'\|_2^2}{4}, \tag{9.27}$$

that is, the extremal $y^* = 0$ is a weak local minimum on the neighborhood $\|h\| < \epsilon_0 = m/4$.

This will be the pattern of what is to come:

To establish that an extremal is a weak local minimum, we verify that the second variation is uniformly positive definite with respect to this weaker Dirichlet norm $\| \cdot \|_2$ and then apply the estimate (9.20).

Finally, because J is a convex functional (Exercise 9.18) on the (convex) space V of all pairs $v = (y, y')$ where y is continuously differentiable on $[0, b]$ with $y(0) = 0 = y(b)$, the weak local minimum $y^* = 0$ is a global minimum (Exercise 1.27).

Example 2. (CVP 7, §3.7) Mechanical motions of the spring/mass system with $k = m = 1$ of Figure 3.7 must, by Hamilton's principle, yield stationary values of the integral

$$J = \int_{t_1}^{t_2} \dot{q}^2 - q^2 \, dt. \tag{9.28}$$

The second variation of this problem is given by an integral of a constant-coefficient quadratic form, namely,

$$\delta^2 J[h] = \int_{t_1}^{t_2} \dot{\eta}^2 - \eta^2 \, dt. \tag{9.29}$$

At first glance, since the integrand of (9.29) is of mixed signature, it seems clear that the motions are neither minima nor maxima of the integral (9.28) of the Lagrangian, as we have verified in Exercises 3.20–3.22. But by Exercise 9.17, when $\eta(t_1) = 0$,

$$\int_{t_1}^{t_2} \eta^2(t) \, dt \leq \frac{(t_2 - t_1)^2}{2} \int_{t_1}^{t_2} \dot{\eta}^2(t) \, dt, \tag{9.30}$$

giving that
$$\delta^2 J[h] \geq \left(1 - \frac{(t_2-t_1)^2}{2}\right) \|\eta'\|_2^2, \quad (9.31)$$

yielding a uniformly positive definite second H-variation for all small time spans with respect to the Dirichlet norm $\|\cdot\|_2$. Thus by the estimate of Lemma A, the mechanical motions are weak local minima during small durations of time provided the initial configuration $q(t_1)$ is specified (see Exercise 3.21).

9.4 The Integrand $p\eta'^2 + q\eta^2$

As is so often the case in mathematical investigations, it all comes down to one special case at the core of the issue.

Question. Let H be the space of all pairs $h = (\eta, \eta')$ where η is absolutely continuous with bounded derivative η' and where $\eta(a) = 0 = \eta(b)$. What conditions guarantee that the quadratic functional

$$Q[h] = \int_a^b p(x)\eta'^2(x) + q(x)\eta^2(x)\, dx \quad (9.32)$$

is positive semidefinite on H? We assume that q is at the very least continuous on $[a,b]$ and that $p > 0$ and continuously differentiable on $[a,b]$; see Exercise 9.26.

Because $\eta(a) = 0 = \eta(b)$, by integrating by parts we may complete the square with a differentiable function w as follows:

$$Q[h] = \int_a^b p\eta'^2 + q\eta^2\, dx = \int_a^b p\eta'^2 + q\eta^2 + (w\eta^2)'\, dx$$
$$= \int_a^b p\left(\eta'^2 + 2\eta'\eta\frac{w}{p} + \frac{w'+q}{p}\eta^2\right) dx = \int_a^b p\left(\eta' + w\eta/p\right)^2 dx \geq 0,$$
$$(9.33)$$

provided w solves on $[a,b]$ the *Riccati* equation

$$w' + q = w^2/p. \quad (9.34)$$

The equation (9.34) is certainly locally solvable by Picard's theorem (Appendix B) but may have a barrier to a continuation to a solution on all of $[a,b]$. See Exercise 9.24.

9.4 The Integrand $p\eta'^2 + q\eta^2$

Fact A. The Riccati equation $(w' + q)p = w^2$ possesses a solution on all of $[a, b]$ if and only if there exists a nonvanishing solution u on all of $[a, b]$ to the linear *Jacobi (accessory, subsidiary)* equation

$$(pu')' = qu. \tag{9.35}$$

Proof. Exercises 9.22 and 9.23.

A practical method for deciding when the accessory equation (9.35) has nonvanishing solutions is provided by the following notion and its accompanying fact.

Definition. Let u be the *principal* solution of $(pu')' = qu$ with $u(a) = 0$ and $u'(a) = 1$. If for some $a^* > a$, $u(a^*) = 0$, then a^* is called a *conjugate point (kinetic focus)* to a of the accessory equation (9.35).

Fact B. The accessory equation $(pu')' = qu$ possesses a nonvanishing solution on $[a, b]$ if and only if there are no points a^* conjugate to a in $(a, b]$.

Proof. Suppose v is any solution and u the principal solution with $u(a) = 0, u'(a) = 1$. Note that the Wronskian

$$W(x) = \det \begin{bmatrix} v(x) & u(x) \\ v'(x) & u'(x) \end{bmatrix}$$

is of constant sign; in fact, $W = K \exp(-p(x))$ since (exercise)

$$W' = -\frac{p'}{p}W.$$

Suppose v is nonvanishing (say) positive on $[a, b]$, yet there are points a^* conjugate to a in $(a, b]$; let a^* be the leftmost such point. Then since u is at first positive, $u'(a^*) \leq 0$. But then, since $W(a) = v(a) > 0$,

$$0 < W(a^*) = \det \begin{bmatrix} v(a^*) & u(a^*) \\ v'(a^*) & u'(a^*) \end{bmatrix}$$

$$= \det \begin{bmatrix} v(a^*) & 0 \\ v'(a^*) & u'(a^*) \end{bmatrix} = v(a^*)u'(a^*) \leq 0,$$

a contradiction.[3]

Conversely, suppose there are no points conjugate to a in $(a, b]$, that is, $u(x) > 0$ on $(a, b]$. Let v be any solution with $v'(a) = 1$ and $v(a) > 0$. If v vanishes on $[a, b]$, say at $x = x_0$, then

$$0 < W(x_0) = v(a) = -u(x_0)v'(x_0), \tag{9.36}$$

and so $v'(x_0) < 0$. Thus v is thereafter negative since the graph of $y = v(x)$ at zero crossings must have negative slope.

On the other hand, solutions exhibit continuous dependence on initial conditions; in fact there exists a constant k for all v with $v'(a) = 1$ so that (Exercise 9.29)

$$|v(x) - u(x)| \leq k|v(a)|. \tag{9.37}$$

Thus for all small positive choices for the value $v(a)$,

$$v(b) > \frac{u(b)}{2} > 0. \tag{9.38}$$

Therefore all such v with sufficiently small positive initial value $v(a)$ are nonvanishing on $[a, b]$.

Lemma B. If there are no points conjugate to a on $(a, b]$, then the quadratic functional (9.32) is uniformly positive definite with respect to the Dirichlet norm, that is,

$$Q[h] = \int_a^b p\eta'^2 + q\eta \, dx \geq \gamma \|\eta'\|_2^2 \tag{9.39}$$

for some positive γ.

Proof. Suppose there are no points in $(a, b]$ conjugate to a. Then by Facts A and B, the Riccati equation $p(w' + q) = w^2$ is solvable on all of $[a, b]$, and so by completing the square,

$$Q[h] = \int_a^b p\eta'^2 + q\eta^2 \, dx = \int_a^b p\left(\eta' + w\eta/p\right)^2 \, dx \geq 0. \tag{9.40}$$

---

[3] This and the following are instances of the rule that the zeros of functions with Wronskian of constant sign must interlace. See Exercise 9.42.

9.5 Weak Local Sufficiency

Note that

$$Q[h] - \gamma \int_a^b \eta'^2 \, dx = \int_a^b (p - \gamma)\eta'^2 + q\eta^2 \, dx \qquad (9.41)$$

for $0 < \gamma$. The accessory equation for this form (9.41) is

$$((p - \gamma)u')' - qu = 0, \qquad (9.42)$$

which for all small $\gamma > 0$ also possesses a nonvanishing solution (Exercise 9.25). Thus $Q[h] - \gamma \|\eta'\|_2^2 \geq 0$ on H.

Example 2. (revisited) The second variation (9.29) of the spring-mass system of Figure 3.7 with $k = m = 1$ is

$$\delta^2 J = \int_{t_1}^{t_2} \eta'^2 - \eta^2 \, dx \qquad (9.43)$$

with accessory equation

$$\ddot{u} = -u \qquad (9.44a)$$

and principal solution

$$u = \sin(t - t_1). \qquad (9.44b)$$

There are clearly no conjugate points in any interval $(t_1, t_2]$ as long as $t_2 - t_1 < \pi$. Thus by Result B, each mechanical motion is a weak local minimum for such time intervals as long as we prescribe the initial and terminal configurations [i.e., the admissible variations η must have $\eta(t_1) = 0 = \eta(t_2)$]. This improves on our previous result of $t_2 - t_1 < \sqrt{2}$ in Example 2 [which assumed only $\eta(t_1) = 0$]. Exercise 3.21 shows this stronger result to be sharp, namely, for time intervals longer than π, the mechanical motions are no longer weak local minima, even when initial and terminal configurations are specified.

9.5 Weak Local Sufficiency

We can now realize our goal of a sufficient condition for weak local minima of costs

$$J = \int_a^b L(x, y(x), y'(x)) \, dx \qquad (9.45)$$

for a certain large class of problems.

Assumptions:[4]

A1. The Lagrangian L is everywhere thrice continuously differentiable.

A2. The path $y = y^*(x)$ on $[a, b]$ is an extremal of the cost (9.45) and belongs to the class C of all continuously differentiable paths $y = y(x)$ on $[a, b]$ with specified initial and terminal values $y(a) = A$ and $y(b) = B$.

A3. $p(x) = L_{y'y'}(x, y^*(x), y^{*'}(x))$ is positive and continuously differentiable on $[a, b]$.

A4. $q(x) = L_{yy}(x, y^*(x), y^{*'}(x)) - \frac{d}{dx}L_{yy'}(x, y^*(x), y^{*'}(x))$ is continuous on $[a, b]$.

A5. There are no points conjugate to a in $(a, b]$ of the accessory equation $(pu')' = qu$.

Theorem. (weak local sufficiency) Under the assumptions A1–A5, the extremal y^* is an isolated local weak minimum of the cost (9.45) within the class C of all continuously differentiable functions y with specified end values $y(a) = A$ and $y(b) = B$.

Proof. Let H be the space of all pairs $h = (\eta, \eta')$ where η is continuously differentiable and $\eta(a) = 0 = \eta(b)$. Thus thinking of C more precisely as the class of all pairs (y, y') with y continuously differentiable and $y(a) = A$, $y(b) = B$, we have $C = (y^*, y^{*'}) + H$.

By parts,

$$2!\delta^2 J[h] = \int_a^b L_{yy}\eta^2 + 2L_{yy'}\eta\eta' + L_{y'y'}\eta'^2 \, dx$$

$$= \int_a^b L_{yy}\eta^2 + 2L_{yy'}\eta\eta' + L_{y'y'}\eta'^2 - (\eta^2 L_{yy'})' \, dx$$

$$= \int_a^b L_{yy}\eta^2 + (L_{yy} - \frac{d}{dx}L_{yy'})\eta^2 \, dx = \int_a^b p\eta'^2 + q\eta^2 \, dx. \quad (9.46)$$

---
[4] Assumption 3 is called the *strong Legendre condition;* see Exercise 9.26. Actually, A3 and A4 follow from A1, A2, and $p > 0$ by a result of Hilbert to be presented in Chapter 11.

9.5 Weak Local Sufficiency

Since there are no points in $(a, b]$ conjugate to a, we have by (9.40) Lemma B that

$$\int_a^b p\eta'^2 + q\eta^2 \, dx \geq \gamma \|\eta'\|_2^2 \qquad (9.47)$$

for some $\gamma > 0$, i.e., the second variation of this extremal y^* is uniformly positive definite with respect to the norm $\| \cdot \|_2$. Therefore, by the estimate (9.20) of Lemma A, for $v_0 = (y^*, y^{*\prime})$,

$$J[v_0 + h] - J[v_0] = \delta^2 J[h] + E[h] \geq \left(\frac{\gamma}{2!} - m\|h\|_0\right) \|\eta'\|_2^2 > 0 \quad (9.48)$$

for all $h \neq 0$ in H in the neighborhood $\|h\|_0 < \epsilon_0 = \gamma/(2m)$. Thus y^* is an isolated minimum.

Example 3. Consider the problem of minimizing the cost

$$J = \int_0^1 xy' + y'^2 \, dx \qquad (9.49)$$

subject to the end conditions $y(0) = \alpha$ and $y(1) = \beta$. Here $L_y = 0$ and $L_{y'y'} = 2$, so the second variation becomes

$$\delta^2 J[h] = \int_0^1 \eta'^2 \, dx, \qquad (9.50)$$

with accessory equation
$$u'' = 0 \qquad (9.51a)$$
and principal solution
$$u = x. \qquad (9.51b)$$

Since there are no points conjugate to 0 in $(0, 1]$, the extremal $y = -x^2/4 + (\beta + 1/4)x + \alpha$ is a weak local minimum.

Example 4. The catenary (CVP 3, §4.2) and the minimal surface CVP 13 (§6.1) were discovered by minimizing a cost

$$J = \int_{-a}^{a} y\sqrt{1 + y'^2} \, dx, \qquad (9.52)$$

yielding extremals

$$y = c \cosh \frac{x}{c} \qquad (9.53)$$

satisfying $y(-a) = A = y(a)$. There are actually two such extremals satisfying these end conditions for all $A/a > \sinh z_0 \approx 1.58882$, where $1 = z_0 \tanh z_0$. See Exercise 9.35.

The second variation for these extremals is (Exercise 9.36)

$$\delta^2 J = \frac{1}{2!} \int_{-a}^{a} L_{y'y'}\eta'^2 + \left(L_{yy} - \frac{d}{dx}L_{yy'}\right)\eta^2 \, dx$$

$$= \frac{1}{2!} \int_{-a}^{a} \frac{y\eta'^2}{(1+y'^2)^{3/2}} - \left(\frac{d}{dx}\frac{y'}{\sqrt{1+y'^2}}\right)\eta^2 \, dx \quad (9.54)$$

$$= \frac{1}{2!} \int_{-a}^{a} \frac{c\eta'^2}{\cosh^2 x/c} - \left(\frac{d}{dx}\frac{\sinh x/c}{\cosh x/c}\right)\eta^2 \, dx \quad (9.55)$$

$$= \frac{1}{2!} \int_{-a}^{a} \frac{c\eta'^2}{\cosh^2 x/c} - \frac{\eta^2}{c\cosh^2 x/c} \, dx. \quad (9.56)$$

The accessory equation is then

$$0 = (pu')' - qu = \left(\frac{cu'}{\cosh^2 x/c}\right)' + \frac{u}{c\cosh^2 x/c},$$

that is,

$$u'' - 2\frac{u'}{c}\tanh\frac{x}{c} + \frac{u}{c^2} = 0. \quad (9.57)$$

Suppose u is a positive solution of (9.57) on $[-a, a]$. Because $v(x) = u(-x)$ is also a solution, we may assume u even with $u'(0) = 0$. Then by numerics (Exercise 9.40) we find that $u(x) > 0$ only for $|x/c| < z_0 \approx 1.19968$. Thus by Exercise 9.35, the upper catenary is a weak local minimum. We will see in Chapter 10 compelling geometric reasons why this must be the case and why the lower catenary is not a weak minimum.

These results are not germane to the original problem of the hanging chain (CVP 3), where competing curves must all have the identical arclength.

Remark. The requirement in Assumption A2 that the class C of admissible curves be continuously differentiable can often be weakened to a wider subclass of the absolutely continuous functions with bounded derivative, say to the class of all piecewise continuously differentiable functions. In many problems, the second partials of the

Exercises

Lagrangian are so simple (e.g., constant) that little or no regularity on the extremal is required for A3 and A4 to hold and thus for the sufficiency theorem to obtain.

We are close to providing conditions that are both necessary and sufficient for a weak local minimum; see Exercise 9.41 and Chapter 11. All of the theoretical results of this chapter easily generalize to multiple degrees of freedom—see [Gelfand and Fomin].

EXERCISES

9.1 Suppose C is the class of all realvalued continuously differentiable functions on $[0, 1]$. Display a sequence $\|\eta_n\|_0 \longrightarrow 0$ of variations of the zero function that are not weak variations.

Suggestion: Try $\eta_n(x) = (1/n) \sin n\pi x$.

9.2 An *inner product* on the real vector space V is a symmetric, positive definite bilinear form on V, that is to say, a functional $\langle . \, , \, . \rangle : V \times V \longrightarrow \mathbb{R}$, satisfying

(i) $\langle u, v \rangle = \langle v, u \rangle$ for all $u, v \in V$

(ii) $\langle cu, v \rangle = c \langle u, v \rangle$ for all $u, v \in V$ and $c \in \mathbb{R}$

(iii) $\langle u + v, w \rangle = \langle u, w \rangle + \langle v, w \rangle$ for all $u, v, w \in V$

(iv) $\langle v, v \rangle \geq 0$ for all $v \in V$, with equality exactly when $v = 0$.

Prove that the **Cauchy-Schwarz inequality** obtains:
$$\langle u, v \rangle \leq \|u\| \cdot \|v\|,$$
where $\|v\| = \sqrt{\langle v, v \rangle}$, with equality exactly when u, v are linearly dependent. Also deduce that $\| \, . \, \|$ is a norm on V.

Hint: Expand out $0 \leq \langle u + cv, u + cv \rangle$; then set $c = \|u\|/\|v\|$.

9.3 Prove that the first variation A of (9.11a) (if extant) is a uniquely determined linear functional.

9.4 Prove that the second variation Q of (9.12a) given by $Q[h] = B[h, h]$ (if extant) is a uniquely determined (quadratic) functional.

9.5 Verify the formulas of (9.19).

9.6 Verify the inequality (9.20).

9.7 Show that without loss of generality we may assume that the matrix A of a quadratic form $q(x) = x^\top A x$ is symmetric.
Hint: $A = (A + A^\top)/2 + (A - A^\top)/2$.

9.8 Show that a quadratic form $q(x) = \langle Ax, x \rangle = x^\top A x$ is a convex function when A is positive semidefinite. (This is a reprise of Exercise 3.15.)

9.9 The integral of a convex functional is a convex functional. In particular, show that if at each x, the Lagrangian $L[y] = L(x, y, y')$ is a convex function of all pairs $v = (y, y')$ drawn from a class C, then the cost functional

$$J[v] = \int_a^b L(x, y(x), y'(x))\, dx$$

is a convex function on C.

9.10 The *signature* of a quadratic form $q(x) = x^\top A x$, where A is a $n \times n$ symmetric matrix, is the sign pattern (r, s, t) of the eigenvalues of A, i.e., r, s, t is the number of positive, negative, and zero eigenvalues respectively, where $n = r + s + t$. Show via the spectral theorem that the form Q is

> positive semidefinite exactly when $s = 0$,
> positive definite exactly when $s = t = 0$,
> negative semidefinite exactly when $r = 0$,
> negative definite exactly when $r = t = 0$.

9.11 We may easily hand compute the signature of a quadratic form via *Sylvester's law of inertia*—the signature of a form is invariant under *congruences,* to whit, a change of basis $x = Py$ transforms the form $q(x) = x^\top A x$ to the form $q_1(y) = y^\top P^\top A P y$, which takes on the identical values once proper identifications are made. By hand, we perform an elementary row operation on A followed by the identical column operation on the result.

Exercises

Show that by a sequence of such pairs of elementary steps, the original form q can be brought to *congruence canonical form*:
$$y^\top P^\top A P y = y_1^2 + \cdots + y_r^2 - y_{r+1}^2 - \cdots - y_{r+s}^2.$$

9.12 Deduce from the diagonalization method of Exercise 9.11 that a quadratic form $q(x) = x^\top A x$ given by a $n \times n$ symmetric A is positive definite if and only if each descending diagonal minor determinant
$$d_k = \det[a_{ij}]_{1 \leq i,j \leq k}, \qquad k = 1, 2, \ldots, n,$$
is positive.

9.13 Hand compute the signature of the quadratic form given by
$$A = \begin{bmatrix} 6 & 3 & 0 \\ 3 & 8 & -3 \\ 0 & -3 & -2 \end{bmatrix}.$$
Answer: $r = 3, s = t = 0$.

9.14 Show that the signature of the quadratic form
$$q(x) = ax_1^2 + 2bx_1 x_2 + cx_2^2$$
is positive definite when $a > 0$ and $(ac - b^2) > 0$.

9.15 Show that a positive definite quadratic form $q(x) = x^\top A x$, A symmetric, is uniformly positive definite. In fact,
$$x^\top A x \geq \gamma |x|^2 = x^\top x,$$
where γ is the smallest positive eigenvalue of A.

9.16 Show that when the absolutely continuous η with bounded derivative satisfies $\eta(a) = 0$, then
$$\|\eta\|_0 = \sup_{a \leq x \leq b} |\eta(x)| \leq \sqrt{b-a} \left(\int_a^b \eta'(x)^2 \, dx \right)^{1/2}.$$

Hint: By Cauchy-Schwarz,
$$\eta(x) = \int_a^x \eta'(t) \, dt \leq \left(\int_a^x 1^2 \, dt \right)^{1/2} \left(\int_a^x \eta'(t)^2 \, dt \right)^{1/2}.$$

9.17 For η absolutely continuous with bounded derivative and $\eta(a) = 0$, obtain the inequality

$$\int_a^b \eta(x)^2 \, dx \le \frac{(b-a)^2}{2} \int_a^b \eta'(x)^2 \, dx.$$

9.18 Show that the arclength functional (9.24) is convex on the (convex) space V of all pairs $v = (y, y')$ where y is continuously differentiable on $[0, b]$ and where $y(0) = 0 = y(b)$.

Hint: Employ Exercises 1.33 or 9.9.

9.19 Show the extremals of mechanical motion of the spring-mass system (Example 2) can never be local minima or maxima when end conditions are not specified.

Hint: Use $\eta(t) = \epsilon \sin(t - \phi)$.

9.20 Let H be the space of all pairs $h = (\eta, \eta')$, where η is an absolutely continuous realvalued function on $[a, b]$ with bounded derivative η' and where $\eta(a) = 0$. Consider the four norms on H given by $\|\eta'\|_0, \|\eta'\|_2, \|\eta\|_0$, and $\|\eta\|_2$. Show that the first dominates the second, which in turn dominates the third which dominates the fourth. (One norm *dominates* a second when every sequence that converges with respect to the first converges with respect to the second.)

9.21 Show that no inequality of the form

$$\int_a^b \eta'^2 \, dx \ge c\|\eta'\|_0 = \sup_{a \le x \le b} |\eta'(x)|$$

is possible for some fixed $c > 0$, even when $\eta(a) = 0 = \eta(b)$. Thus the Dirichlet norm $\|\eta'\|_2$ is weaker than the supremum norm $\|\eta'\|_0$.

Hint: Consider $\eta(x) = x - x^n$ on $[0, 1]$.

9.22 Prove that the Riccati equation $p(w' + q) = w^2$ has a solution on all of $[a, b]$ if and only if there exists a solution of the Jacobi equation $(pu')' = qu$ that vanishes nowhere on $[a, b]$.

Outline: If there is a solution w of the Riccati equation, set $u = \exp(-\int_a^x w(t) \, dt/p(t))$ to obtain a nonvanishing solution to $(pu')' = qu$. Conversely, if u is a nonvanishing solution to $(pu')' = qu$, set $w = -pu'/u$ to obtain a solution to $w'+q = w^2$.

9.23 Prove that the linear ODE $(pu')' = qu$ does not suffer barriers to continuation—any solution can be extended to any larger interval upon which q is continuous, and p continuously differentiable and positive.

Hint: Reread the proof of Picard's theorem (Appendix B) very carefully. The δ can be uniformly chosen for linear ODEs.

9.24 In contrast to Exercise 9.23, find the barrier to continuing any solution of the Riccati equation $w' - 1 = w^2$ about $x = 0$ to (say) $[-\pi, \pi]$.

9.25 Under the assumptions of Lemma B, prove that if $(pu')' = qu$ has no points conjugate to a in $(a, b]$, then neither does $((p - \gamma)u')' - qu = 0$ for all small γ.

Outline: Let u be a positive solution to $(pu')' = qu$ and v the solution to $(p - \gamma)v')' = qv$ with the identical initial values. Use the "E-trick" from the proof of the perturbation lemma of Appendix B to show that the solutions v are uniformly bounded for all small γ. Then apply the perturbation lemma.

9.26 Prove *Legendre's condition:* For an extremal to be a weak local minimum, it is necessary that $L_{y'y'} \geq 0$ along the extremal path.

Outline: If an extremal is minimal, the second variation must be positive semidefinite. Rewrite $2!\delta^2 J[h]$ to have integrand $p\eta'^2 + q\eta^2$. Suppose $p = L_{y'y'} < -\alpha < 0$ on $I = (x_0 - \epsilon, x_0 + \epsilon)$, where ϵ is small enough for the estimate (9.22) to hold over this $2\epsilon_0$ interval I of integration. Choose a variation η with small derivative that vanishes off I yet is positive on I.

9.27 Show that the strong Legendre condition $L_{y'y'} > 0$ implies that the second variation (9.19d) is always uniformly positive definite for sufficiently small integration intervals $[a, b]$. Deduce the following:

The strong Legendre condition guarantees that extremals are weak local minima over small intervals.

9.28 Show that the extremal (4.14) of CVP 2, the graph with least surface area of revolution, is a weak minimum.

9.29 Verify that the inequality (9.37) is correct.

Hint: Go to state space and apply the perturbation lemma of Appendix B.

9.30 Find the continuously differentiable weak local minimum of

$$J = \int_0^1 xy' + y'^2 \, dx$$

subject to $y(0) = 1$, $y(1) = 0$.

9.31 Find the continuously differentiable weak local minimum of

$$J = \int_0^1 y'^2 + 2y'y - 4y^2 \, dx$$

subject to $y(0) = 0$, $y(1) = 1$.

9.32 Find a continuously differentiable weak local minimum of

$$J = \int_a^b y\sqrt{1 + y'^2} \, dx$$

subject to $y(a) = \cosh a$, $y(b) = \cosh b$, $0 < a < b$.

Hint: $u = \sinh x$ is a positive solution of (9.57) on $(0, \infty)$ when $c = 1$.

9.33 Find the continuously differentiable weak local minimum of the cost

$$J = \int_1^2 y' + x^2 y'^2 \, dx$$

subject to $y(1) = 1, y(2) = 2$.

9.34 Find a continuously differentiable weak local minimum of the cost

$$J = \int_0^b y^2 + y'^2 \, dx$$

subject to $y(0) = 0, y(b) = B$.

9.35 First, show that the catenaries $y = c \cosh x/c$ all share a certain tangent line $y = mx$ for all $c > 0$, where $m = \sinh z_0 \approx 1.508882$, $1 = z_0 \tanh z_0$, $z_0 \approx 1.19968$.

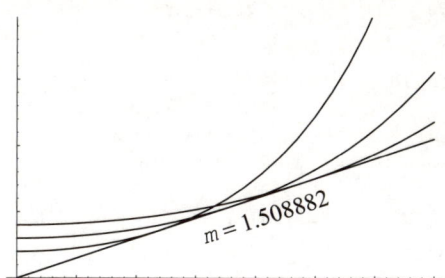

Second, show that given two points $(\pm a, A)$ with $a, A > 0$ within the interior of the cone bounded by $y = \pm mx$, then there are exactly two such catenaries passing through the two given points, only one if the two points lie on the boundary, and none if the points are exterior.

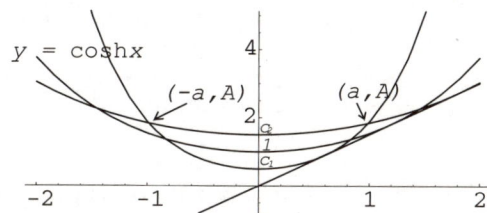

9.36 Verify (9.54)–(9.56).

9.37 Test the extremals for weak local minima or maxima of the cost
$$J = \int_0^1 y'^2 (1 + y'^2)^2 \, dx,$$
where $y(0) = 0, y(1) = \beta$.

9.38* Prove that as normed-space-to-normed-space functionals, the first variation of the Euler-Lagrange functional (at an extremal) is the Jacobi functional, that is,
$$\delta \left(\frac{d}{dx} L_{y'} - L_y \right) [h] = (p\eta')' - q\eta,$$
where $h = (\eta, \eta')$, $p = L_{y'y'}$, and $q = L_{yy} - (L_{yy'})'$.

Hint: Employ (9.16a).

9.39 Show that the Euler-Lagrange equation of the second variation
$$\delta^2 J = \frac{1}{2!} \int_a^b p\eta'^2 + q\eta^2 \, dx$$
is the Jacobi equation $(p\eta')' - q\eta = 0$.

9.40 Show numerically that any even positive solution of Jacobi equation (9.57) must vanish at $x/c \approx 1.19968$.

Outline: By means of a substitution, we may assume $c = 1$. Employ the *Mathematica* script
```
sol = NDSolve[{u''[x]-2*Tanh[x]*u'[x]+u[x]==0,u[0]==1,u'[0]==0},
u,{x,0,1.2}]; f[x_] := u[x] /. sol[[1]]; Plot[f[x],{x,0,1.2}];
```

9.41 Prove that if the second variation (9.19d) is positive definite for all piecewise continuously differentiable η with $\eta(a) = 0 = \eta(b)$, then the Jacobi equation (9.60) has no points conjugate to a in $(a, b]$.

Hint: Set $\eta(x) = u(x)$ on $[a, a^*]$, zero otherwise. Integrate $\delta^2 J$ by parts.

9.42 Is the cycloid (4.25) a weak local minimum of the brachistochrone problem?

9.43 Suppose (i) u and v are continuously differentiable, (ii) their Wronskian
$$W(x) = \det \begin{bmatrix} u(x) & v(x) \\ u'(x) & v'(x) \end{bmatrix}$$
is of constant sign on $\mathbb{R}$, and (iii) they share no zeros. Prove that their respective zeros must interlace—each zero of u must lie between two zeros of v and *vice versa*.

Deduce that no nonzero solution of (9.57) can have two positive zeros.

Hint: $y = c \sinh(x/c)$ is a solution.

9.44 Can there be a candidate extremal of the form
$$y = c \cosh \frac{x+d}{c}$$
distinct from $y = \cosh x$ for the problem of Exercise 9.32?

Chapter 10
Strong Sufficiency

Knowing that an extremal is a weak minimum is not always satisfactory (or even safe). There may be an extremal nearby in value but not in derivative with better (safer) performance. Our next task will be to obtain a guarantee for when an extremal is in fact a strong minimum or maximum. To this end we first review the notion of *flows* of second-order ODEs, then specialize to flows of the Euler-Lagrange equation. An analog of one of Hamilton's equations, called the *Hamilton-Jacobi* equation, plus a miraculous artifice of Hilbert, simplifies the criterion for strong minima to the sign of the *Weierstrass E-function*. We work many examples to demonstrate the practicality of the approach and then conclude with a theoretical investigation into the existence of flows.

10.1 The Goal

Our objective is to obtain a practical test for when a twice continuously differentiable extremal $y = y^*(x)$ of the cost functional

$$J[y] = \int_a^b L(x, y, y')\, dx \qquad (10.1a)$$

subject to the end conditions

$$y(a) = \alpha, \text{ and } y(b) = \beta \qquad (10.1b)$$

is in fact a strong local minimum. That is, when can we be sure for all twice continuously differentiable y satisfying the end conditions that we have

$$J[y] \geq J[y^*] \qquad (10.2)$$

on some neighborhood

$$\|y - y^*\|_0 = \sup_{a \leq x \leq b} |y(x) - y^*(x)| < \epsilon? \qquad (10.3)$$

Since such strong local minima are weak local minima, the obvious first thought is to search for a fifth assumption, which, when combined with the four assumptions of §9.5 guaranteeing weak minima, will guarantee strong minima. However, we will see that there is a much simpler test for practical problems. But we must first review several notions from elementary ODEs.

10.2 Flows

Consider the family of solutions $y = y(x)$ on $[a, b]$ to the second-order ODE
$$y'' = F(x, y, y') \tag{10.4}$$
with F continuous. The solution graphs form hopeless tangles (see Figure 10.1). But by Picard (Appendix B), we may pick out a unique solution curve $y = y(x)$ from this family by specifying two initial conditions $(y(a), y'(a))$, or, for that matter, by specifying intermediate conditions $(y(x_0), y'(x_0))$ for any $a \leq x_0 \leq b$.

Figure 10.1 suggests that solutions to (10.4) come in bundles of parallel or at least nonintersecting curves, that is, *streamlines* forming a *flow* (of solutions).

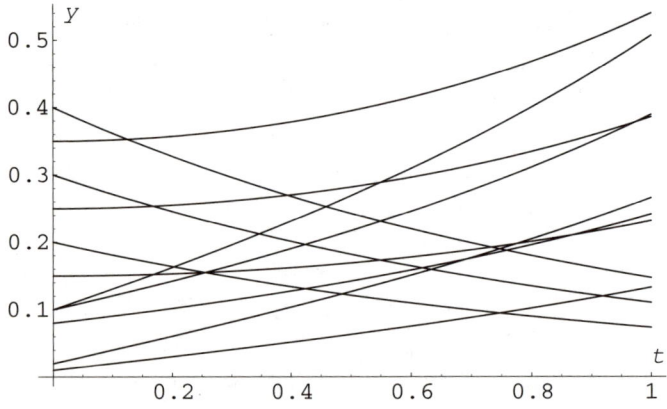

Figure 10.1 The tangle of solutions to $y'' = y$.

10.2 Flows

Definition. A function $\psi = \psi(x, y)$ continuously differentiable on a simply connected[1] open set $\Omega \subset \mathbb{R} \times \mathbb{R}$ is a *flow* of the second-order ODE (10.4) if every *streamline* $y = y(x)$ in Ω, that is, solution of

$$y' = \psi(x, y) \qquad (10.5)$$

is a solution of (10.4). (Flows are also called *direction fields*.)

Remark A. Streamlines of the same flow are nonintersecting, for again by Picard, solution paths of the first-order equation $y' = \psi(x, y)$ cannot meet.

Example 1. Consider the family of solutions of
$$y'' = y, \qquad (10.6)$$
namely $y = a \cosh x + b \sinh x$. Figure 10.1 displays the tangle of some of these solutions. We may select the flow of nonintersecting solutions consisting of all $y = a \cosh x$ by observing that $y' = a \sinh x = a \tanh x \cosh x = y \tanh x$. This suggests that we choose

$$\psi(x, y) = y \tanh x, \qquad \Omega = \mathbb{R} \times \mathbb{R}. \qquad (10.7)$$

Note that every solution (streamline) of $y' = \psi(x, y)$ is a solution of (10.6) (Exercise 10.4) and so ψ is a flow of (10.6).

In order to visualize a flow, we graph short line segments of slope $m = \psi(x, y(x))$ centered at $(x, y(x))$ for selected points $(x, y(x))$. A simple script will reveal qualitatively the flow of (10.7):

```
% Routine 10.1 Direction Fields (Mathematica)
<<Graphics`PlotField`
a = 0; b = 1; min = 0; max = 1;
psi[x_,y_] := y*Tanh[x]
PlotVectorField[{ 1, psi[x,y]},{x,a,b},{y,min,max}]
```

which returns with Figure 10.2. The streamlines must have these line segments as tangents.

Or, for another example, what flow of nonintersecting solutions of $y'' = y$ contains the streamline $y = \cosh x - \sinh x = e^{-x}$? Since $y' = -e^{-x} = -y$, we see that this solution belongs to the flow

$$y' = \psi(x, y) = -y, \qquad \Omega = \mathbb{R} \times \mathbb{R}, \qquad (10.8a)$$

---

[1] An open set Ω is *simply connected* if it is arcwise connected and if every simple closed curve in Ω contains no points not in Ω, i.e., Ω has no 'holes.'

yielding the flow of nonintersecting streamlines

$$y = ke^{-x}. \qquad (10.8b)$$

Figure 10.2 The flow of solutions to $y'' = y$ given by $\psi = y \tanh x$.

Alert 1. Flows need not be disjoint. The same solution curve of (10.4) may belong to distinct flows ψ_1, ψ_2. See Exercise 10.1.

Returning to the general problem (10.4), wherever a flow ψ is defined on an open neighborhood, we may differentiate (10.5) via the chain rule to discover that ψ satisfies the *Hamilton-Jacobi* partial differential equation (Exercise 10.3)

$$\frac{\partial \psi}{\partial x} + \psi \frac{\partial \psi}{\partial y} = F(x, y, \psi). \qquad (10.9)$$

Conversely, given any solution $\psi = \psi(x, y)$ to the Hamilton-Jacobi PDE (10.9) on an open set O, we may construct (at least local) solutions to (10.4) merely by solving locally the first-order ODE

$$y' = \psi(x, y). \qquad (10.10)$$

More ambitiously, suppose we can solve the Hamilton-Jacobi equation (10.9) for ψ on an open, simply connected set Ω that contains both (a, α) and (b, β). Moreover, suppose we may then solve

$$y' = \psi(x, y) \qquad (10.11)$$

to obtain a solution $y = y(x)$ with a graph in Ω that connects the end conditions (a, α) and (b, β) as in Figure 10.3. Then $y = y(x)$ solves the boundary value problem

$$y'' = F(x, y, y') \quad \text{subject to } y(a) = \alpha \text{ and } y(b) = \beta. \qquad (10.12)$$

10.2 Flows

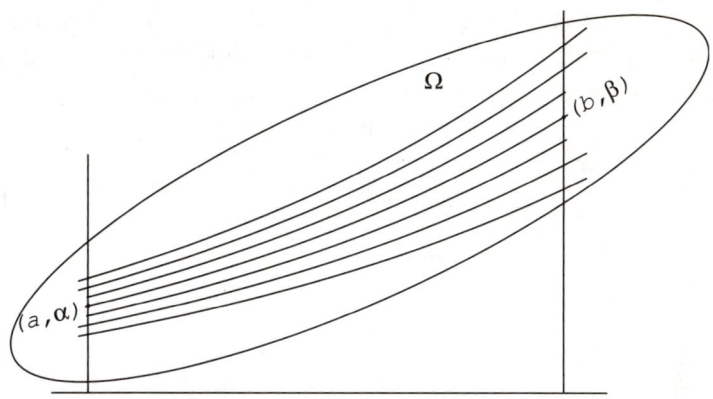

Figure 10.3 Several streamlines of a flow $y' = \psi(x,y)$ of $y'' = F(x,y,y')$ on the open simply connected domain Ω. One of the streamlines y solves the boundary value problem $y'' = F$ subject to $y(a) = \alpha$ and $y(b) = \beta$.

We will use later the startling fact that for any streamline Γ of the flow ψ and for any continuous $G(x,y)$, the line integral

$$I = \int_\Gamma \psi G \, dx - G \, dy = 0, \tag{10.13}$$

since along Γ, $dy = \psi(x, y(x)) \, dx$.

Summary: A continuously differentiable function $\psi = \psi(x,y)$ defined on some open simply connected set Ω is a *flow* of the second-order ODE (10.4) if every solution of (10.5) is a solution of (10.4). Each flow of (10.4) satisfies the Hamilton-Jacobi partial differential equation (10.9), and any continuously differentiable solution of (10.9) on an simply connected open set Ω is a flow of (10.4).

10.3 Flows of the Euler-Lagrange Equation

Standing Assumptions: Assume $y = y^*(x)$ is a twice differentiable extremal of the cost functional

$$J[y] = \int_a^b L(x, y, y') \, dx, \qquad (10.14)$$

satisfying the end conditions $y(a) = \alpha$ and $y(b) = \beta$. Assume the Lagrangian L is thrice continuously differentiable on $\Omega \times (-\infty, \infty)$, where Ω is an open simply connected set containing the extremal path $y = y^*(x)$ for $a \leq x \leq b$. Assume $L_{y'y'}(x, y^*(x), y^{*\prime}(x)) > 0$ on $[a, b]$. Last, assume

$$\psi = \psi(x, y) \qquad (10.15)$$

is a flow of the Euler-Lagrange equation

$$\frac{d}{dx} L_{y'} - L_y = y'' L_{y'y'} + y' L_{y'y} + L_{y'x} - L_y = 0 \qquad (10.16)$$

defined on Ω that contains y^* as one of its streamlines.

Using a mechanical metaphor, we define on $\Omega \times \psi(\Omega)$ the *generalized momentum*

$$\tilde{p} = L_{y'}(x, y, \psi) \qquad (10.17)$$

and the *Hamiltonian*

$$\tilde{H} = \tilde{p}\psi - L(x, y, \psi). \qquad (10.18)$$

(Compare with (4.38) and (4.39).)

Lemma A. Analogously to Hamilton's equations (4.41), we have on Ω the *Hamilton-Jacobi* equation

$$\frac{\partial \tilde{p}(x, y, \psi(x, y))}{\partial x} = -\frac{\partial \tilde{H}(x, y, \psi(x, y))}{\partial y}. \qquad (10.19)$$

Proof. By careful applications of the chain rule,

$$\tilde{p}_x + \tilde{H}_y = L_{y'x} + L_{y'y'}\psi_x + \left(L_{y'y} + L_{y'y'}\psi_y\right)\psi + L_{y'}\psi_y - L_y - L_{y'}\psi_y$$

$$= (\psi_x + \psi\psi_y) L_{y'y'} + \psi L_{y'y} + L_{y'x} - L_y.$$

10.3 Flows of the Euler-Lagrange Equation

But every point of Ω lies on exactly one streamline $y' = \psi(x,y)$ of the flow ψ of (10.16). Therefore,

$$\tilde{p}_x + \tilde{H}_y = y''L_{y'y'} + y'L_{y'y} + L_{y'x} - L_y = \frac{d}{dx}L_{y'} - L_y = 0.$$

Corollary A. The differential form

$$-\tilde{H}(x,y,\psi(x,y))\,dx + \tilde{p}(x,y,\psi(x,y))\,dy \qquad (10.20)$$

is an exact differential on Ω.

Proof. Since the cross partials are equal and the domain Ω is simply connected, Stokes's theorem [Folland] guarantees the existence of a potential.

Corollary B. The value of the *Hilbert invariant line integral*

$$I_\Gamma = \int_\Gamma -\tilde{H}\,dx + \tilde{p}\,dy \qquad (10.21)$$

with $\Gamma \subset \Omega$ is independent of the path Γ as long as the endpoints of Γ are fixed.

Corollary C. When the path Γ is the extremal path $y = y^*(x)$, the Hilbert integral (10.21) devolves to the cost $J[y^*]$, that is,

$$I_\Gamma = \int_a^b L(x, y^*, y^{*\prime})\,dx = J[y^*].$$

Proof.

$$\int_\Gamma -\tilde{H}\,dx + \tilde{p}\,dy = \int_\Gamma (L - \psi\tilde{p})\,dx + \tilde{p}\,dy$$

$$= \int_\Gamma L\,dx + \int_\Gamma -\tilde{p}\psi\,dx + \tilde{p}\,dy = \int_\Gamma L\,dx + \int_\Gamma -\tilde{p}\psi\,dx + \tilde{p}\,\psi\,dx$$

$$= \int_\Gamma L\,dx = \int_a^b L\,dx.$$

Corollary D. The cost $J[y^*]$ at the extremal y^* coincides with the Hilbert integral (10.21) along *any* path Γ in Ω that begins at (a, α) and ends at (b, β).

Lemma B. Let $\Gamma : y = y(x)$ be any piecewise continuously differentiable path in Ω that begins at (a, α) and ends at (b, β). Then

$$J[y] - J[y^*] = \int_a^b E(x, y, y^{*\prime}, y') \, dx, \qquad (10.22)$$

where E is the *Weierstrass excess function*

$$E(x, y, z, w) = L(x, y, w) - L(x, y, z) - (w - z)L_{y'}(x, y, z). \qquad (10.23)$$

Proof. By Corollary D,

$$J[y] - J[y^*] = \int_\Gamma L \, dx - \int_\Gamma -\tilde{H} \, dx + \tilde{p} \, dy = \int_\Gamma \left(L + \tilde{H} \right) dx - \tilde{p} \, dy$$

$$= \int_\Gamma L(x, y, y') - L(x, y, \psi) + \tilde{p}(x, y, \psi)\psi - \tilde{p}(x, y, \psi)y' \, dx$$

$$= \int_a^b E(x, y, \psi, y') \, dx. \qquad (10.24)$$

10.4 The E-Function and Strong Sufficiency

Under the standing assumptions and notation in force in the preceding section, we finally have, via Lemma B, a sufficient condition that will guarantee that extremals are strong local minima.

Theorem A. (local strong sufficiency) If the Weierstrass E-function (10.23) is nonnegative for all 4-tuples (x, y, z, w) where $(x, y) \in \Omega$, $z \in \psi(\Omega)$, and $w \in \mathbb{R}$, then the extremal y^* is a strong local minimum of the cost (10.14).

Example 2. Let us test the extremal $y^* = 0$ to the arclength integral of CVP 1:

$$J[y] = \int_0^b \sqrt{1 + y'^2} \, dx, \quad y(0) = 0 = y(b). \qquad (10.25)$$

The Euler-Lagrange equation for this Lagrangian is

$$y'' = 0,$$

10.4 The E-Function and Strong Sufficiency

which clearly possesses the flow

$$\psi(x, y) \equiv 0, \quad \Omega = \mathbb{R} \times \mathbb{R},$$

with $y^* = 0$ as one streamline. Because the Lagrangian $L(x, y, z) = \sqrt{1 + z^2}$, the Weierstrass E-function is

$$E(x, y, z, w) = L(x, y, w) - L(x, y, z) - (w - z)L_{y'}(x, y, z)$$

$$= \sqrt{1 + w^2} - \sqrt{1 + z^2} - \frac{(w - z)z}{\sqrt{1 + z^2}}$$

$$= \frac{\sqrt{1 + z^2}\sqrt{1 + w^2} - zw - 1}{\sqrt{1 + z^2}}, \qquad (10.26)$$

which by Exercise 10.6 is nonnegative. Thus the extremal $y^* = 0$ is a (global) minimum since the domain of the flow $\Omega = \mathbb{R} \times \mathbb{R}$ places no restriction on the strong neighborhoods $\|y - y^*\|_0 < \epsilon$.

Example 3. (Exercise 9.33 reprised) Consider the minimization of

$$J[y] = \int_1^2 y' + x^2 y'^2 \, dx$$

subject to $y(1) = 1$ and $y(2) = 2$. The Euler-Lagrange equation for this cost is

$$0 = x^2 y'' + 2xy' = (x^2 y')',$$

yielding the unique candidate extremal

$$y^*(x) = -2/x + 3.$$

This extremal is a streamline of the flow

$$\psi(x, y) = 2/x^2, \quad \Omega = (0, \infty) \times \mathbb{R},$$

which is in fact a flow of the Euler-Lagrange equation.

Because $L(x, y, z) = z + x^2 z^2$, the Weierstrass E-function of the cost is

$$E(x, y, z, w) = w + x^2 w^2 - z - x^2 z^2 - (w - z)(1 + 2x^2 z) = x^2(w - z)^2 \geq 0,$$

and thus the extremal is a minimum. (The verification of all these details is Exercise 10.7.)

Alert 2. The preceding trick of setting the flow ψ equal to the derivative of the candidate extremal y^* rarely works—extraneous streamlines usually appear. A flow certainly results, but not necessarily a flow of the intended second-order equation. See Exercise 10.5. We must check carefully that each streamline of $y' = \psi(x, y)$ solves the Euler-Lagrange equation of the given cost functional $J[y]$.

Example 4. (Exercise 9.34 reprised) Consider the minimization of the cost
$$J[y] = \int_0^b y^2 + y'^2 \, dx$$
subject to the end conditions $y(0) = 1$ and $y(b) = B > 1$. The Euler-Lagrange equation for this cost is
$$y'' - y = 0$$
with the unique candidate extremal
$$y^*(x) = \frac{\cosh x}{B},$$
which is a streamline of the flow
$$\psi(x, y) = y \tanh x, \qquad \Omega = \mathbb{R} \times \mathbb{R}.$$
This flow is a flow of the Euler-Lagrange equation $y'' = y$. Because the Lagrangian $L(x, y, z) = y^2 + z^2$, the Weierstrass E-function is
$$E = y^2 + w^2 - y^2 - z^2 - (w - z)2z = (w - z)^2 \geq 0,$$
and so the extremal y^* is a minimum. (The verification of all these details is Exercise 10.8.)

Example 5. Let us again revisit the cost that gives rise to the catenary (CVP 3, CVP 13):
$$J[y] = \int_a^b y\sqrt{1 + y'^2} \, dx,$$
but in this iteration subject to the very special end conditions $y(a) = \cosh a$, $y(b) = \cosh b$, $0 < a < b$. The Euler-Lagrange equation is
$$yy'' = 1 + y'^2,$$

10.4 The E-Function and Strong Sufficiency

with a candidate extremal

$$y = \cosh x$$

and flow

$$\psi(x,y) = \sqrt{y^2 - 1}, \quad \Omega = (0, \infty) \times (1, \infty).$$

The Weierstrass E-function is

$$E(x, y, z, w) = y\sqrt{1 + w^2} - y\sqrt{1 + z^2} - (w - z)\frac{yz}{\sqrt{1 + z^2}}$$

$$= y\frac{\sqrt{1 + z^2}\sqrt{1 + w^2} - zw - 1}{\sqrt{1 + z^2}}.$$

Hence $E \geq 0$ when $y > 0$. Thus the extremal $y = \cosh x$ is a strong local minimum. (The verification of all these details is Exercise 10.9.)

Example 6. Let us revisit the spring-mass system CVP 7 with cost

$$J[q] = \int_{t_1}^{t_2} \dot{q}^2 - q^2 \, dt$$

subject to $q(t_1) = q_1$, $q(t_2) = q_2$, and Lagrange equation

$$\ddot{q} = -q.$$

Since any two motions $q = q(t)$ meet when $t_2 - t_1 \geq \pi$, it is impossible to construct a flow ψ of nonintersecting streamlines for such durations.

On the other hand, when $t_2 - t_1 < \pi$, we will see in the next section that a flow $\psi(t, q)$ exists on a certain domain $\Omega = (t_1 - \epsilon, t_2 + \epsilon) \times (-a, a)$, and thus the (unique) extremal is a strong local minimum (Exercise 10.11).

Remark B. Our theorem does not exploit the full power of (10.22). Lemma B only requires $E(x, y, y^{*\prime}(x), w) \geq 0$, while Theorem A requires the stronger assumption $E(x, y, z, w) \geq 0$.

10.5 Existence of Flows

Suppose we have a candidate extremal y^* with a nonnegative Weierstrass E-function. We suspect it is a strong local minimum but have been unable to find a flow with y^* as streamline. Is there a theoretical tool for asserting the existence of such a flow without having to display it explictly?

Again it is very helpful to think more generally.

Lemma C. Suppose $x = x^*(t)$ is a solution of the first-order vector boundary value problem
$$\dot{x} = F(x) \tag{10.27a}$$
satisfying
$$x^*(a) = \alpha \text{ and } x^*(b) = \beta, \tag{10.27b}$$
where F is continuously differentiable on some open domain of $\mathbb{R}^n$ containing the path $\Gamma : x = x^*(t)$ for $a \leq t \leq b$. Then there exists an open neighborhood V of β and an open neighborhood U of α such that each point y^0 in V is the terminal point of a solution of (10.27a) that initiates at $t = a$ from a point[2] x^0 of U. (See Figure 10.4.)

Proof sketch. We may assume that the lemma holds for all terminal points $x^*(t)$ along the path Γ as long as $t < b$. Find neighborhoods of β and of b where Picard's theorem holds.

Consider the flow
$$x(t) = X(x^0, t),$$
that is, the solution[3] of (10.27a) initiating at x^0 that must (by Picard) exist for $|t - b| \leq \delta$ and all x^0 close to β.

Consider the map
$$G(x^0) = X(x^0, b - \delta). \tag{10.28}$$

Note that G is a map from a neighborhood of β back in time to previous points, and, in particular, $G(\beta) = x^*(b-\delta)$. Once we check that the Jacobian $G'(\beta)$ is nonsingular, the inverse function theorem (Appendix A) guarantees that G is a continuously differentiable bijective map from an open neighborhood of β to an open neighborhood of $x^*(b - \delta_0)$, a neighborhood that, when cut back, is covered

---
[2] The point x^0 is unique if the duration of the trip is considered.
[3] The object $X(x^0, t)$ is the more standard notion of a *flow*.

10.5 Existence of Flows

by solutions initiating near α. Continue these solutions by reversing the flow $X(x^0, b-t)$ to cover a neighborhood of β.

To establish the nonsingularity of G, we exploit the fixed point statement:
$$X(x^0, t) = x^0 + \int_0^t F(X(x^0, s)) \, ds. \tag{10.29}$$

The difference quotient for the partial derivatives with respect to x_j^0 is then
$$\frac{X(x^0 + he^j, t) - X(x^0, t)}{h}$$
$$= e^j + \frac{1}{h} \int_0^t F(X(x^0 + he^j, s)) - F(X(x^0, s)) \, ds$$
$$= e^j + \int_0^t F'(\xi(x^0, s)) \frac{X(x^0 + he^j, t) - X(x^0, t)}{h} \, ds$$
$$= e^j + \int_0^t F'(X(x^0, s)) \frac{X(x^0 + he^j, t) - X(x^0, t)}{h} \, ds + o(1), \tag{10.30}$$

where it is first necessary to check that the difference quotients are bounded—apply the "E-trick" of Appendix B. Thus
$$\frac{X(x^0 + he^j, t) - X(x^0, t)}{h} = (I - A_{x^0})^{-1} e^j + o(1), \tag{10.31}$$

where A_{x^0} is the contractive linear map
$$(A_{x^0} f)(t) = \int_a^t F'(X(x^0, s)) f(s) \, ds.$$

Therefore, in the limit,
$$G'(\beta) = \left[\frac{\partial X(\beta, \delta)_i}{\partial x_j} \right] = (I - A_\beta)^{-1}.$$

Corollary. There is an open flow tube of nonintersecting solutions of (10.27a) containing the curve Γ. (See Figure 10.4.)

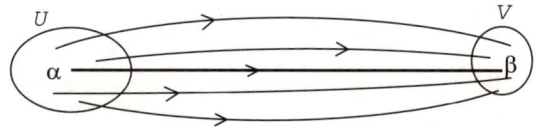

Figure 10.4 An open tube of nonintersecting solutions of the autonomous $dx/dt = F(x)$ about one solution.

Revised Standing Assumptions. Assume $y = y^*(x)$ is a twice differentiable extremal of the cost functional

$$J[y] = \int_a^b L(x, y, y') \, dx, \qquad (10.32)$$

satisfying the end conditions $y(a) = \alpha$ and $y(b) = \beta$. Assume the Lagrangian $L(x, y, z)$ is thrice continuously differentiable on $\Omega \times (-B, B)$, where Ω is an open simply connected subset of the plane containing the path Γ of $y = y^*(x)$. Assume

$$L_{y'y'}(x, y^*(x), y^{*\prime}(x)) > 0 \qquad (10.33)$$

on $[a, b]$. Assume that the Weierstrass E-function

$$E(x, y, z, w) = L(x, y, w) - L(x, y, z) - (w - z)L_y(x, y, z) \qquad (10.34)$$

is nonnegative on $\Omega \times (-B, B) \times (\infty, \infty)$. Finally, assume that the Jacobi accessory equation

$$(pu')' - qu = (L_{y'y'}u')' - (L_{yy} - (L_{y'y})')u = 0 \qquad (10.35)$$

has no points conjugate to a in $(a, b]$.

By our theorem of Chapter 9, the extremal y^* is a weak local minimum. But is it a strong local minimum? We will now verify in six steps the existence of a flow $y' = \psi(x, y)$ of the Euler-Lagrange equation with y^* as a streamline. Thus by our strong sufficiency Theorem A of this chapter, our extremal y^* is indeed a strong local minimum.

First, write the Euler-Lagrange equation as

$$y'' = F(x, y, y')$$
$$= -\frac{y' L_{y'y}(x, y, y') + L_{y'x}(x, y, y') - L_y(x, y, y')}{L_{y'y'}(x, y, y')}. \qquad (10.36)$$

Note that the Euler-Lagrange equation as an autonomous system is then given by

$$\begin{vmatrix} y \\ y' \\ x \end{vmatrix}' = \begin{vmatrix} y' \\ F(x, y, y') \\ 1 \end{vmatrix}. \qquad (10.37)$$

10.5 Existence of Flows

Second, by Picard, we may extend the extremal y^* to an extremal on some larger interval $[a - \delta, b + \delta]$ with $y^*(a - \delta) = \alpha_0$. Choose $\delta > 0$ small enough that the Jacobi equation (10.35) has no points conjugate to $a - \delta$ on $(a - \delta, b + \delta]$ (Exercise 10.20). By Lemma C applied to (10.37) there exists a *central field* $\mathcal{F}$ of extremals $y = y(x)$ initiating at α_0 at $x = a - \delta$ covering a neighborhood of $\beta_0 = y^*(a+\delta)$, achieved by shooting for β_0 with all initial slopes y' close to $y^{*\prime}(a-\delta)$. The central field $\mathcal{F}$ forms an open tube surrounding our extremal path $\Gamma : y = y^*(x)$ for $a - \delta \leq x \leq b + \delta$. Smaller intervals of initial slopes yield tubes of smaller distance from Γ. See Figure 10.5.

Does this central field $\mathcal{F}$ of extremals contain a flow $y' = \psi(x, y)$ of nonintersecting extremals on $[a, b]$?

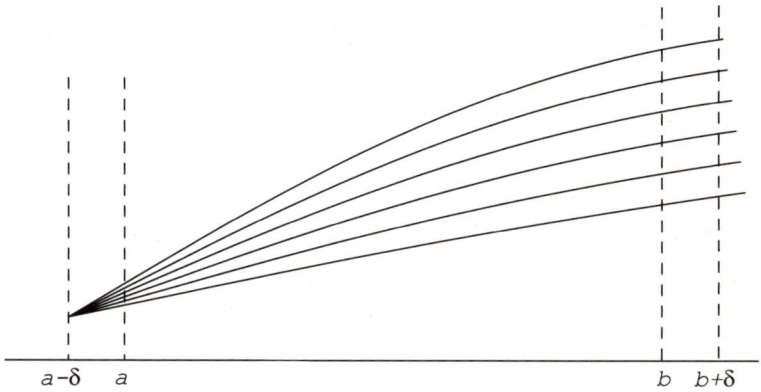

Figure 10.5 The one-parameter central field of extremals emanating from $(a-\delta, \alpha_0)$.

Third, let $y = y(x, \zeta)$ denote the extremal in $\mathcal{F}$ launched by the initial slope ζ at $x = a - \delta$, i.e., $\zeta = y'(a - \delta, \zeta)$. Of course, $y(a - \delta, \zeta) = \alpha_0$. Moreover, revisiting Exercise 9.38, we see that (Exercise 10.21)

$$
\begin{aligned}
0 &= \frac{\partial}{\partial \zeta} \left[\frac{d}{dx} L_{y'}(x, y(x, \zeta), y'(x, \zeta)) - L_y(x, y(x, \zeta), y'(x, \zeta)) \right] \\
&= \frac{d}{dx} \left[\frac{\partial}{\partial \zeta} L_{y'}(x, y(x, \zeta), y'(x, \zeta)) \right] - \frac{\partial}{\partial \zeta} L_y(x, y(x, \zeta), y'(x, \zeta)) \\
&= \frac{d}{dx} \left[L_{y'y} y_\zeta + L_{y'y'} y'_\zeta \right] - L_{yy} y_\zeta - L_{yy'} y'_\zeta \\
&= \left(L_{y'y'} (y_\zeta)' \right)' - \left(L_{yy} - (L_{yy'})' \right) y_\zeta. \quad (10.38)
\end{aligned}
$$

Thus
$$u = \frac{\partial}{\partial \zeta} y(x, y^{*\prime}(a - \delta)) \qquad (10.39)$$

satisfies Jacobi's equation (10.35) on $[a - \delta, b + \delta]$ with $u(a - \delta) = 0$. If necessary, decrease δ so that u is not identically 0.

Fourth, fix a point $(x_0, y^*(x_0))$ on the extremal path Γ, x_0 in $[a, b]$. Consider for each small neighborhood $|\zeta - y^{*\prime}(a-\delta)| < \epsilon_1$ an open ball $B(\epsilon_1)$ centered at $(x_0, y^*(x_0))$ consisting of points on the extremals $y(x, \zeta)$. Suppose each such $B(\epsilon_1)$ contains a point of intersection of at least two of these extremals. Then $(x_0, y^*(x_0))$ is a cluster point of such intersections and hence $u(x_0) = y_\zeta(x_0, y^{*\prime}(x_0)) = 0$, which contradicts the nonexistence of points conjugate to $a - \delta$ in $(a - \delta, b + \delta]$. Thus there are no intersections within all small balls $B(\epsilon_1)$. This is an instance of an interesting geometric fact:

A point p^* is conjugate to a point p if p^* is a cluster point of abscissae of intersections of a central field of extremals that all emanate from (p, P).

Fifth, cover the extremal path $\Gamma : y = y^*(x)$, $a \leq x \leq b$, with these open neighborhoods $B(\epsilon_1)$ consisting of points on extremals $y = y(x, \zeta)$ that do not intersect. Extract a finite subcover, then construct an open simply connected domain Ω containing Γ consisting of extremals from $y = y(x, \zeta)$ launched from some small neighborhood of initial slopes: $|\zeta - y^{*\prime}(a - \delta)| < \epsilon_0$.

Finally, since each point (x, y) in Ω is on exactly one extremal $y = y(x)$ from $\mathcal{F}$, this central field $\mathcal{F}$ forms a flow $y' = \psi(x, y)$ on Ω for the Euler-Lagrange equation (10.36) on $[a, b]$ (Exercise 10.22), which by construction has $y = y^*(x)$ embedded as a streamline.

Thus we have our second sufficiency theorem.

Theorem B. Under the revised standing assumptions of this section, the extremal $y = y^*(x)$ is a strong local minimum of the cost (10.32).

Example 5. (reprise) Let us again examine the catenary problem CVP 3—the minimization of

$$J[y] = \int_{-a}^{a} y\sqrt{1 + y'^2} \, dx \qquad (10.40)$$

subject to $y(-a) = A = y(a)$. The Euler-Lagrange equation is

10.5 Existence of Flows

$$yy'' = 1 + y'^2, \tag{10.41}$$

with first integral ("no-x")

$$y^2 = c^2(1 + y'^2), \tag{10.42}$$

with solutions

$$y = c \cosh \frac{x - \phi}{c},$$

each of which is a solution of the Euler-Lagrange equation (10.41).

The candidate extremal satisfying the end conditions is therefore of the form

$$y = c \cosh \frac{x}{c},$$

with

$$\frac{A}{c} = \cosh \frac{a}{c}.$$

The excess function

$$E(x, y, z, w) = y \frac{\sqrt{1 + z^2}\sqrt{1 + w^2} - zw - 1}{\sqrt{1 + z^2}} \geq 0$$

for $y > 0$.

Reviewing the results of Exercise 9.35, for the end conditions to hold, $(\pm a, A)$ must lie within the cone bounded by the tangent lines $y = \pm mx$ shared by all $y = c \cosh(x/c)$, where $m = \sinh z_0 \approx 1.508882$, where $1 = z_0 \tanh z_0$, and where $z_0 \approx 1.19968$. If (a, A) lies on the boundary of the cone, there is exactly one extremal; if within the interior, two through $(\pm a, A)$. See Figure 10.6.

Assume $(\pm a, A)$ is within the interior, (i.e., $A/a > m = \sinh z_0$). The upper catenary with $c > 1$ is obtained by dilating $y = \cosh x$ by c, the second by contracting $y = \cosh x$ by $c < 1$.

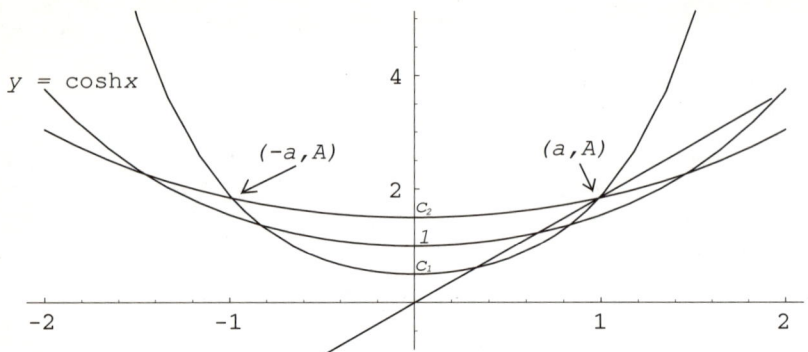

Figure 10.6 The two candidate extremals $y = c_i \cosh(x/c_i)$ through the points $(-a,A)$ and (a,A). The upper catenary with $c_2 > 1$ is obtained by dilating the curve $y = \cosh x$ by c_2, while the lower is obtained by a contraction by c_1.

But in fact, $(-a, A)$ is the center of a field of extremals of (10.41) emanating from $(-a, A)$ obtained by varying c and ϕ in tandem, that is, we may solve for c in terms of ϕ by means of

$$A - c \cosh \frac{a + \phi}{c} = 0 \tag{10.43}$$

via the implicit function theorem[4] (Appendix A) to obtain a one-parameter family $\mathcal{F}$ of curves of the form

$$y = c(\phi) \cosh \frac{x - \phi}{c(\phi)} \tag{10.44}$$

through $(-a, A)$ as in Figure 10.7. And, as an instance of the third step of the proof of Theorem B,

$$u = \frac{\partial y}{\partial \phi} = c_\phi(0) \left[\cosh \frac{x}{c} - \frac{x}{c} \sinh \frac{x}{c} \right] - \sinh \frac{x}{c} \tag{10.45}$$

solves (Exercise 10.28) the Jacobi accessory equation

$$u'' - 2 \frac{u'}{c} \tanh \frac{x - \phi}{c} + \frac{u}{c^2} = 0, \tag{10.46}$$

where $c = c(0)$ and $c_\phi = \partial c / \partial \phi$.

---

[4] See Exercise 10.30.

10.5 Existence of Flows

Note that (Exercise 10.29)

$$u(x) = \cosh\frac{x}{c} - \frac{x}{c}\sinh\frac{x}{c} \qquad (10.47a)$$

is an even, positive solution of the accessory equation (10.46) for

$$-cz_0 < x < cz_0. \qquad (10.47b)$$

Therefore, by Theorem B, since there is no point conjugate to $-a$ in $(-a, a)$ when $a < cz_0$,

the upper catenary is a strong local minimum.

(This case is shown in Figure 10.7.) However, as we will prove in Chapter 11, *Jacobi's necessary condition* for weak local minima guarantees that when the strong Legendre condition $p > 0$ is present, the interior of the interval over which an extremal is a weak local minimum cannot possess conjugate points. Hence

the lower catenary is not a strong local minimum

since $(-a, a)$ contains the two conjugate points $\pm cz_0$.

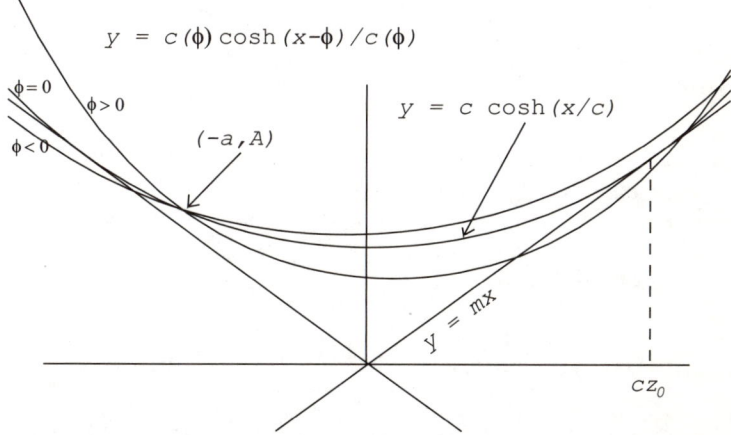

Figure 10.7 By solving for c as a function of ϕ, we construct a central field of extremals emanating from $(-a, A)$ whose intersections with the candidate extremal cluster at its point of tangency.

Thinking more geometrically, as $\phi \to 0$, the intersections of the corresponding extrema of (10.44) with the extremal $y = c\cosh(x/c)$ approach (cz_0, mcz_0), and hence cz_0 must be a point conjugate to

$-a$. See Figure 10.7 (where only the upper catenary case is shown). The lower catenary thus has two conjugate points in $(-a, a)$, while the upper has none.

One case remains—where $(\pm a, A)$ lies on the boundary of the cone bounded by $y = \pm mx$. We may assume by similarity that $c = 1$, $a = z_0$, so that the candidate extremal is $y = \cosh x$. Because

$$\eta(x) = \cosh x - x \sinh x \tag{10.48}$$

vanishes at the endpoints $\pm a$, we may use it as a variation of our candidate extremal $y^*(x) = \cosh x$ to obtain the perturbed cost

$$J[y^* + \epsilon\eta] = J[y^*] + \epsilon\delta J[\eta] + \epsilon^2 \delta^2 J[\eta] + \epsilon^3 \delta^3 J[\eta] + \epsilon^4 E_4. \tag{10.49}$$

Since y^* is an extremal, the first variation vanishes. Because η satisfies the accessory equation, the second variation vanishes at η since by parts,

$$2!\delta J[\eta] = \int_{-a}^{a} p\eta'^2 + q\eta^2 \, dx = -\int_{-a}^{a} [(p\eta')' - q\eta] \eta \, dx = 0. \tag{10.50}$$

The third variation

$$\delta^3 J[\eta] = \frac{1}{3!} \int_{-a}^{a} L_{y'y'y'}\eta'^3 + 3L_{y'y'y}\eta'^2\eta + 3L_{y'yy}\eta'\eta^2 + L_{yyy}\eta^3 \, dx$$

$$= \text{(Exercise 10.31)} = \frac{1}{2} \int_{-a}^{a} \frac{-\eta'^3 \sinh x}{\cosh^4 x} + \frac{\eta'^2 \eta}{\cosh^3 x} \, dx \tag{10.51}$$

$$= \text{(Exercise 10.32)} = \int_{0}^{a} \frac{\eta'^2}{\cosh^4 x} [1 - \sinh^2 x] \, dx \tag{10.52}$$

$$= \int_{0}^{z_0} \frac{(2\sinh x - x\cosh x)^2 (1 - \sinh^2 x)}{\cosh^4 x} \, dx$$

$$= \text{(Exercise 10.32)} > 0.034. \tag{10.53}$$

Therefore, since the third variation is not 0 at η, the extremal $y = \cosh x$ is not even a local *weak* minimum.

All of the ideas of this chapter generalize easily to multiple-degrees-of-freedom systems with one additional complication—the *self-adjointness* condition on the generalized momenta. See [Gelfand and Fomin].

EXERCISES

10.1 Find a second flow ψ of $y'' = y$, containing the streamline $y = \cosh x$, that is distinct from $\psi = y \tanh x$. Graph streamlines of these two flows.

Sample Answer: $\psi(x,y) = \sqrt{y^2 - 1}$, $\quad \Omega = (0, \infty) \times (1, \infty)$.

10.2 Find a flow ψ of $y'' = y$ containing the streamline $y = 4\cosh x + 2\sinh x$.

Answer: $\psi(x,y) = \sqrt{y^2 - 12}$ on the domain $\Omega = \mathbb{R} \times (\sqrt{12}, \infty)$.

10.3 Verify that flows ψ of (10.4) satisfy the Hamilton-Jacobi equation (10.9).

10.4 Verify that every streamline (solution) of $y' = \psi(x,y) = y \tanh x$ solves (10.6).

10.5 Show that although the flow $y' = \psi(x,y) = \sinh x$ contains the streamline $y = \cosh x$, it is not a flow of $y'' = y$.

10.6 Establish that the excess function (10.26) is everywhere non-negative.

Hint: Set $z = \sinh \theta$ and $w = \sinh \phi$, or apply (1.18).

10.7 Carefully check all details of Example 3.

10.8 Carefully check all details of Example 4.

10.9 Carefully check all details of Example 5.

10.10 Find (if any) the strong local minima of the problem

$$J[y] = \int_{-1}^{1} 12y^2 + x^2 y'^2 \, dx,$$

subject to $y(-1) = -1$, $y(1) = 1$.

10.11 Verify for Example 6 that when $t_2 - t_1 < \pi$, there exists a flow ψ on $\Omega = (t_1 - \epsilon, t_2 + \epsilon) \times (-a, a)$, giving that the (unique) candidate extremal is a strong local minimum.

10.12 [Gelfand and Fomin] Investigate the extrema of the following functionals:

(a) $J[y] = \int_{-1}^{2} y'(1 + x^2 y')\, dx \quad y(-1) = 1,\ y(2) = 1.$

(b) $J[y] = \int_{0}^{\pi/4} 4y^2 - y'^2 + 8y\, dx \quad y(0) = -1,\ y(\pi/4) = 0.$

(c) $J[y] = \int_{1}^{2} 12y^2 + x^2 y'^2\, dx \quad y(1) = 1,\ y(2) = 8.$

(d) $J[y] = \int_{0}^{1} y'^2 + y^2 + 2y e^{2x}\, dx, \quad y(0) = 1/3,\ y(1) = e^2/3.$

Answers: In part (b), $y = \sin 2x - 1$ is a strong maximum. In part (d), $y = e^{2x}/3$ is a strong minimum.

10.13 (Project) Delineate the contributions of Jacobi, Weierstrass, Hilbert, and Noether to the miraculous results of this chapter.

10.14 Prove that the Hamilton-Jacobi equation (10.18) is an instance of (10.9).

10.15 Find and classify the extremals of the minimization problem

$$J[y] = \int_{0}^{b} y^2 + y'^2\, dx$$

subject to the end conditions $y(0) = 1$ and $y(b) = B < 1$.

10.16 Prove that the cycloid (4.25) is a strong local minimum of the brachistochrone problem.

Outline: Since the Euler-Lagrange equation possesses the flow

$$\psi(x, y) = \frac{\sqrt{\gamma^2 - y}}{\sqrt{y}}, \quad \Omega = (0, \infty) \times (0, \gamma^2),$$

with the cycloid (4.25) as one streamline, and since

$$E(x, y, z, w) = \frac{1}{\sqrt{y}\sqrt{1 + z^2}} \left[\sqrt{1 + w^2}\sqrt{1 + z^2} - wz - 1\right] \geq 0,$$

we need only fret over the possible singularities at the ends of the trajectory.

Exercises

10.17 Where is the Weierstrass excess function for the functional
$$J[y] = \int_a^b x\sqrt{1+y'^2}\, dx$$
nonnegative?

10.18 Prove *Weierstrass's necessary condition* (under the standing assumptions and notation): If $y = y^*(x)$ is a strong local minimum, then $E(x, y(x), y^{*\prime}(x), w) \geq 0$ for all w and curves $y = y(x)$ with $\|y - y^*\|_0 < \epsilon$, $a \leq x \leq b$.

Outline: Suppose $E(x_0, y(x_0), y^{*\prime}(x_0), w_0) < 0$. Choose y close to y^* everywhere, and in fact equal except on $(x_0 - \epsilon, x_0 + \epsilon)$, where $y'(x_0) = w_0$. By the definition of E and minimality of y^*,
$$0 \leq (J[y] - J[y^*])/\epsilon$$
$$= \frac{1}{\epsilon}\int_{x_0}^{x_0+\epsilon} E(x, y, y^{*\prime}, y')\, dx + \frac{1}{\epsilon}\int_{x_0}^{x_0+\epsilon} (y' - y^{*\prime})L_{y'}(x, y, y^{*\prime})\, dx.$$
Integrate the second integral by parts.

10.19 Carefully work through all details of the proof of Lemma C.

10.20 Prove within the context of §10.5 that if $(pu')' - qu = 0$ has no points conjugate to a in $(a, b]$, then its has no points conjugate to $a - \epsilon$ in $(a - \epsilon, b + \epsilon]$ for all small $\epsilon > 0$.

Hint: See Exercise 9.32.

10.21 Verify (10.38).

10.22 Verify that the extremals from the central field $\mathcal{F}$ forming Ω do indeed form a flow $\psi = \psi(x, y)$ of the Euler-Lagrange equation (10.34). In particular, establish that ψ is continuously differentiable.

10.23 Find the Weierstrass excess function for the functional
$$J[y] = \int_a^b 1 - y'^2\, dx.$$

10.24 Analyze the minimization problem
$$J[y] = \int_1^e xy'^2\, dx, \quad y(1) = 0,\ y(e) = 1.$$

10.25 [Sagan] Find the extremals for the following problems. Are they strong or weak local minima?

(a) $J[y] = \int_0^2 xy' + y'^2 \, dx$, $y(0) = 1$, $y(2) = 0$

(b) $J[y] = \int_0^b y'^2 + 2yy' - 16y^2 \, dx$, $y(0) = 0$, $y(b) = 0$, $b > 0$

(c) $J[y] = \int_0^\pi 4y'^2 - 4y^2 - 8y \, dx$, $y(0) = -1$, $y(\pi/4) = 0$

(d) $J[y] = \int_0^b y'^2 \, dx$, $y(0) = 0$, $b > 0$, $y(b) > 0$.

10.26 Analyze the minimization problem

$$J[y] = \int_0^a y\sqrt{1 + y'^2} \, dx, \quad y(0) = c > 0, \ y(a) = c\cosh(a/c).$$

Hint: $y = c\sinh(x/c)$ is the principal solution of (10.46) on $[0, \infty)$.

10.27 Display two distinct second-order ODEs that share a flow in common.

10.28 Verify by calculation that the partial derivative of y in (10.44) is indeed u given in (10.45) and that this u in turn satisfies the Jacobi equation (10.46).

10.29 Verify by calculation that the u of (10.47a) is an even, positive solution of the accessory equation (10.46) on the domain (10.47b).

10.30 Prove that the hypotheses for the implicit function theorem are met when solving (10.43) for c in terms of ϕ whenever $a \neq cz_0$.

10.31 Verify (10.51).

10.32 Verify (10.52).

10.33 Verify (10.53).

10.34 Why is cz_0 of Example 5 not conjugate to $a = 0$? There is, after all, a pencil of extremals emanating from $(0, c)$.

Chapter 11
Corner Points

In this last short chapter we will discover the surprising regularity shared by extremals when the admissible curves are piecewise smooth. This will also be a careful revisiting of the conditions necessary for minima, both weak and strong. First we define the notion of corner points, then deduce several surprising regularity results for extremals. Next we first deduce Jacobi's necessary condition for weak local minima, then Weierstrass's necessary condition on the excess function for strong local minima and its geometric implication via the figurative. Legendre's necessary condition for weak local minima is a byproduct. We finish with two proofs of Erdmann's second corner condition. Some of what follows is drawn from unpublished elegant class notes of Frank H. Clarke.

11.1 Corners and Extremals

Definition. A function $y = y(x)$ continuous on $[a, b]$ is *piecewise continuously differentiable on* $[a, b]$ if it is continuously differentiable except possibly at a finite number of interior points, where the lefthand and righthand derivatives exist but differ in value.

Definition. A piecewise continuously differentiable path $y = y(x)$ on $[a, b]$ is said to have a *corner* at $a < x_0 < b$ if $y'(x_0^-) \neq y'(x_0^+)$.

We restrict ourselves in this chapter to the *standard problem:* Minimize the cost

$$J[y] = \int_a^b L(x, y, y')\, dx, \tag{11.1a}$$

subject to the end conditions

$$y(a) = A \quad \text{and} \quad y(b) = B \tag{11.1b}$$

over all piecewise continuously differentiable paths $y = y(x)$ on $[a, b]$. The Lagrangian L is assumed to be at the very least continuously differentiable.

Definition. Suppose that $y = y(x)$ is a piecewise continuously differentiable path on $[a, b]$ and that the Lagrangian $L = L(x, y, z)$ is continuously differentiable on an open domain Ω containing all points $(x, y(x), y'(x^-))$ and $(x, y(x), y'(x^+))$ for $a \le x \le b$. Then $y = y(x)$ is said to be an *extremal* of the cost functional (11.1a) if for some constant c (which depends upon the particular extremal)

$$L_{y'}(x, y(x), y'(x)) = c + \int_a^x L_y(x, y(x), y'(x))\, dx \qquad (11.2)$$

at all noncorner points on $[a, b]$. See (4.55) and Exercise 11.19.

Example 1. Consider the optimization problem

$$J[y] = \frac{1}{3} \int_0^2 y'^{\,3} \, dx \qquad (11.3)$$

subject to $y(0) = 0 = y(2)$. The integral form of the Euler-Lagrange equation (11.2) for this problem is

$$y'^{\,2} = c. \qquad (11.4)$$

Clearly the only continuously differentiable extremal satisfying the end conditions is the zero function. But there is an infinite number of piecewise continuously differentiable extremals of the form

$$y_m(x) = \begin{cases} mx & \text{if } 0 \le x \le 1 \\ m(2-x) & \text{if } 1 \le x \le 2 \end{cases}. \qquad (11.5)$$

See Figure 11.1.

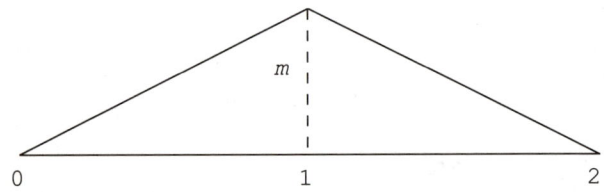

Figure 11.1 There is an infinitude of piecewise continuous extremals for the problem (11.3), each with a corner at $x = 1$.

11.2 First Erdmann Corner Condition

The first extremal regularity result is not surprising and is in fact an obvious corollary of the integral form of the Euler-Lagrange equation (11.2).

Result A. (First Erdmann corner condition) At each corner point x_0 of a piecewise continuously differentiable extremal y of the cost (11.1a),

$$L_{y'}(x_0, y(x_0), y'(x_0^-)) = L_{y'}(x_0, y(x_0), y'(x_0^+)). \tag{11.6}$$

Thus every discontinuity of the function $x \mapsto L_{y'}(x, y(x), y'(x))$ is removable, i.e., $L_{y'}$ is continuous along each extremal.

Example 2. In CVP 1 we were to minimize

$$J[y] = \int_0^b \sqrt{1 + y'^2}\, dx$$

subject to $y(0) = 0 = y(b)$. All extremals must satisfy (11.2) on $[0, b]$, which in this case is

$$\frac{y'}{\sqrt{1 + y'^2}} = c. \tag{11.7}$$

Thus (Exercise 11.1) y' must be continuous, and hence $y = 0$.

In contrast to Result A, the next regularity result is stunning.

Result B. (Hilbert) Suppose the Lagrangian L is twice continuously differentiable, that $y = y(x)$ is a piecewise continuously differentiable extremal of the cost (11.1a), and that

$$p = L_{y'y'}(x_0, y(x_0), y'(x_0)) \neq 0, \tag{11.8}$$

where x_0 is *not* a corner point of $y = y(x)$. Then not only does y'' exist at $x = x_0$, but it is in fact continuous on an open neighborhood of x_0.

Proof. Because L is twice differentiable, by the mean value theorem

$$\frac{L_{y'}(x, y(x), y'(x)) - L_{y'}(x_0, y(x_0), y'(x_0))}{x - x_0}$$

$$= \nabla L_{y'}(x^*, y^*, z^*) \cdot \left(\frac{x - x_0}{x - x_0}, \frac{y(x) - y(x_0)}{x - x_0}, \frac{y'(x) - y'(x_0)}{x - x_0} \right)$$

$$= L_{y'x}(*) + L_{y'y}(*) \frac{y(x) - y(x_0)}{x - x_0} + L_{y'y'}(*) \frac{y'(x) - y'(x_0)}{x - x_0}. \quad (11.9)$$

But by the Euler-Lagrange integral equation (11.2), at noncorner points, $L_{y'}$ is differentiable, in fact,

$$\lim_{x \to x_0} \frac{L_{y'}(x) - L_{y'}(x_0)}{x - x_0} = L_y(x_0), \quad (11.10)$$

where we have abbreviated evaluation at $(x, y(x), y'(x))$ to simply x. But then from (11.9),

$$\lim_{x \to x_0} \frac{y'(x) - y'(x_0)}{x - x_0} = -\frac{y'(x_0) L_{y'y}(x_0) + L_{y'x}(x_0) - L_y(x_0)}{L_{y'y'}(x_0)},$$

and the second-order Euler-Lagrange equation obtains.

Corollary A. Continuously differentiable extremals are twice continuously differentiable wherever $p = L_{y'y'}$ is nonvanishing.

Corollary B. (Jacobi's necessary nondition) Suppose that the continuously differentiable extremal $y = y(x)$ is a weak local minimum of the standard problem (11.1), where the Lagrangian L is thrice continuously differentiable on an open domain Ω containing the path $y = y(x)$. Suppose also that the strong Legendre condition holds on $[a, b]$, that is,

$$L_{y'y'}(x, y(x), y'(x)) > 0, \qquad a \le x \le b. \quad (11.11)$$

Then there are no points conjugate to a in (a, b).

Proof. Suppose there is a point a^* conjugate to a in (a, b), that is, if u is the principal solution to the Jacobi equation[1]

$$(pu')' - qu = 0, \quad (11.12)$$

then $u(a^*) = 0$. Set

$$\eta(x) = \begin{cases} u(x) & \text{if } a \le x \le a^* \\ 0 & \text{otherwise} \end{cases}. \quad (11.13)$$

---

[1] The functions p and q are defined as usual as in (10.35).

Consider the minimization problem

$$I[\eta] = \int_a^b p\eta'^2 + q\eta^2 \, dx \qquad (11.14)$$

over all piecewise continuously differentiable paths subject to $\eta(a) = 0 = \eta(b)$. The Euler-Lagrange equation for problem (11.14) is (11.12), and η of (11.13) is an extremal. By Erdmann's first corner condition,

$$2p\eta' = 2L_{y'y'}(x, y(x), y'(x))\, \eta'(x) \qquad (11.15)$$

must be continuous on $[a, b]$. But because of the strong Legendre condition, η' must then be continuous at the conjugate point a^*. This means the second-order ODE (11.12) satisfies the conditions $u(a^*) = 0 = u'(a^*)$ and must therefore vanish everywhere, contradicting that u is the (nonzero) principal solution of (11.12).

11.3 The Figurative

Suppose the Lagrangian $L = L(x, y, z)$ is continuously differentiable and $y = y(x)$ a piecewise continuously differentiable extremal of the cost (11.1a). The Weierstrass excess function

$$E(x, y, z, w) = L(x, y, w) - L(x, y, z) - (w - z)L_y(x, y, z) \qquad (11.16)$$

has a compelling geometric interpretation.

Definition. The graph of

$$f_x(z) = L(x, y(x), z), \qquad a \le x \le b, \qquad (11.17)$$

is called the *figurative* of the extremal y at the point x.

Lemma A. At each noncorner point x_0, set $y_0 = y(x_0)$ and $z_0 = y'(x_0)$. Then the value

$$E = E(x_0, y_0, z_0, w) \qquad (11.18)$$

of the excess function represents how high the figurative at w lies above its tangent at $z_0 = y'(x_0)$, that is,

$$f_{x_0}(w) = E(x_0, y_0, z_0, w) + f_{x_0}(z_0) + (w - z_0)\frac{d}{dz}f_{x_0}(z_0).$$

214   Chapter 11. Corner Points

See Figure 11.2.

At corner points x_0, the first Erdmann condition guarantees that the tangent lines to the figurative at $z_0^- = y(x_0^-)$ and $z_0^+ = y(x_0^+)$ are parallel.

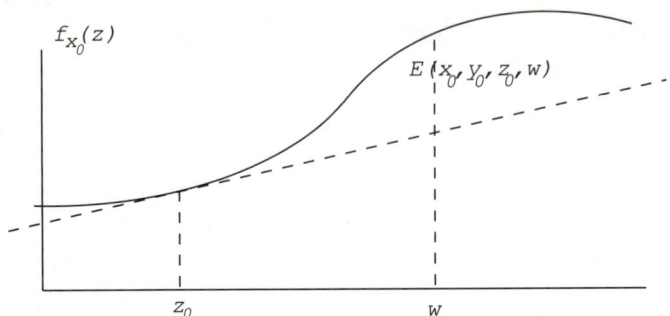

Figure 11.2 The figurative provides a geometric interpretation of the Weierstrass E-function; it is the excess of the Lagrangian L(x,y(x),z) over its tangent line at y'(x).

Result C. (Weierstrass's necessary condition) If $y = y(x)$ is a piecewise continuously differentiable strong local minimum of the standard problem (11.1), then at each noncorner point x_0, the figurative is everywhere above its tangent line at $y'(x_0)$, that is, we have
$$E(x_0, y(x_0), y'(x_0), w) \geq 0 \quad \text{for all } w. \tag{11.19}$$

Proof. Suppose to the contrary that
$$E(x, y(x), y'(x), w_0) < -\alpha < 0 \quad \text{on } I = (x_0 - \epsilon, x_0 + \epsilon). \tag{11.20}$$

Suppose for some k and each small positive ϵ we can find a piecewise continuously differentiable $y = \hat{y}(x)$ that is within $k\epsilon$ of y on the interval $I = (x_0 - \epsilon, x_0 + \epsilon)$ and in fact agrees with y off I, is continuously differentiable on I, but has derivative $\hat{y}'(x_0) = w_0$. Since y is a local strong minimum,
$$0 \leq \frac{J[\hat{y}] - J[y]}{2\epsilon} = \frac{1}{2\epsilon} \int_{x_0 - \epsilon}^{x_0 + \epsilon} L(x, \hat{y}(x), \hat{y}'(x)) - L(x, y(x), y'(x)) \, dx$$
$$= \frac{1}{2\epsilon} \int_{x_0 - \epsilon}^{x_0 + \epsilon} \left[L(x, y, \hat{y}') - L(x, y, y') \right] + \left[L(x, \hat{y}, \hat{y}') - L(x, y, \hat{y}') \right] dx$$

11.3 The Figurative

$$= \frac{1}{2\epsilon} \int_{x_0-\epsilon}^{x_0+\epsilon} L(x,y,\hat{y}') - L(x,y,y') \, dx + o(1)$$

$$= \frac{1}{2\epsilon} \int_{x_0-\epsilon}^{x_0+\epsilon} E(x,y,y',\hat{y}') + (\hat{y}' - y')L_{y'}(x,y,y') \, dx + o(1)$$

$$= \frac{1}{2\epsilon} \int_{x_0-\epsilon}^{x_0+\epsilon} E(x,y,y',\hat{y}') - (\hat{y} - y)\frac{d}{dx}L_{y'}(x,y,y') \, dx + o(1)$$

$$= \frac{1}{2\epsilon} \int_{x_0-\epsilon}^{x_0+\epsilon} E(x,y,y',\hat{y}') - (\hat{y} - y)L_y(x,y,y') \, dx + o(1)$$

$$= \frac{1}{2\epsilon} \int_{x_0-\epsilon}^{x_0+\epsilon} E(x,y,y',\hat{y}') \, dx + o(1) + o(1),$$

which by assumption (11.20) is negative in the limit, a contradiction.

It remains to choose $\hat{y}$: Set $\hat{y} = y + \eta$, where

$$\eta(x) = \begin{cases} \epsilon \frac{w_0 - y'(x_0)}{\pi} \sin \frac{(x-x_0)\pi}{\epsilon} & \text{if } x \in I \\ 0 & \text{otherwise} \end{cases}. \quad (11.21)$$

Then $|\hat{y}(x) - y(x)| \le \epsilon |w_0 - y'(x_0)|/\pi$ and $\hat{y}'(x_0) = w_0$.

Corollary A. The Weierstrass condition (11.19) holds even at corners x_0, that is,

$$E(x_0, y(x_0), y'(x_0^-), w) \ge 0 \text{ and } E(x_0, y(x_0), y'(x_0^+), w) \ge 0. \quad (11.22)$$

Proof. The excess function E is the sum of three terms:

$L(x,y(x),w)$, $-L(x,y(x),y'(x))$, and $-(w-y'(x))L_{y'}(x,y(x),y'(x))$.

The first and second terms, as well as the first factor of the third term, have left limits at corners. The second factor of the third term is by Erdmann continuous along extremals. Therefore, the left limit of E is nonnegative. The same argument applies to right limits.

Corollary B. The Weierstrass condition (11.19) holds at each non-corner point x_0 of a weak local minimum[2] $y = y(x)$ provided w is sufficiently near $y'(x_0)$. Moreover, at corner points x_0, we have

$$E(x_0, y(x_0), y'(x_0^-), w) \ge 0$$

---

[2] When variations η are allowed to possess corner points, we must generalize the norm of (9.6) to $\|\eta\|_0 = \sup_{a \le x \le b} \max(|\eta'(x^-)|, |\eta'(x^+)|)$.

and
$$E(x_0, y(x_0), y'(x_0^+), w) \geq 0$$
for all w sufficiently close to $y'(x_0^-)$ and $y'(x_0^-)$, respectively.

Thus *for weak local minima, the figurative locally lies above its tangent.*

Proof. From (11.21),
$$\|\hat{y}' - y'\|_0 \leq |w_0 - y'(x_0)|.$$

Thus the proof of Result C requires only a slight modification in order to apply to weak minima (Exercise 11.3).

We may deduce that the corollary holds as well for w near $y(x_0^-)$ and $y(x_0^+)$ at corners x_0 via the continuity arguments of the proof of Corollary A.

Corollary C. (Legendre's necessary condition) If $y = y(x)$ is a weak local minimum and the Lagrangian L is twice continuously differentiable, then
$$L_{y'y'}(x, y(x), y'(x)) \geq 0. \tag{11.23}$$

Proof. Fix x at a noncorner point. Expanding in a Taylor series, we obtain
$$L(x, y, w) = L(x, y, y') + (w - y')L_{y'}(x, y, y') + \frac{(w - y')^2}{2!} L_{y'y'}(x, y, v), \tag{11.24}$$
where v lies between w and y'. Thus by Corollary B, for w close to y',
$$L_{y'y'}(x, y, v) = \frac{2!}{(w - y')^2} E(x, y, y', w) \geq 0, \tag{11.25}$$

an inequality that must also hold in the limit as $w \to y'$. The same argument goes through when y' is replaced by either $y'(x^-)$ or $y'(x^+)$.

11.4 Second Erdmann Corner Condition

Result D. (Second Erdmann corner condition) If $y = y(x)$ is a piecewise continuously differentiable strong local minimum of the standard problem (11.1), then at each corner x_0,

$$y'(x_0^-)L_{y'}(x_0, y(x_0), y'(x_0^-)) - L(x_0, y(x_0), y'(x_0^-))$$
$$= y'(x_0^+)L_{y'}(x_0, y(x_0), y'(x_0^+)) - L(x_0, y(x_0), y'(x_0^+)). \quad (11.26)$$

That is, all discontinuities of $x \mapsto L - y'L_{y'}$ are removable. More informally, $L - y'L_{y'}$ is continuous along $y = y(x)$.

First Proof. By Corollary A of the Weierstrass condition (Result C),

$$E(x_0, y(x_0), y'(x_0^-), y'(x_0^+)) \geq 0 \quad (11.27\text{a})$$

and

$$E(x_0, y(x_0), y'(x_0^+), y'(x_0^-)) \geq 0. \quad (11.27\text{b})$$

But by the first Erdmann condition,

$$E(x_0, y(x_0), y'(x_0^-), y'(x_0^+)) = -E(x_0, y(x_0), y'(x_0^+), y'(x_0^-)). \quad (11.28)$$

Hence

$$E(x_0, y(x_0), y'(x_0^-), y'(x_0^+)) = 0, \quad (11.29)$$

which is (11.26) in disguise.

The geometric interpretation of this second corner condition is as follows (Exercise 11.4):

The two tangent lines to the figurative at $z_0^- = y'(x_0^-)$ and $z_0^+ = y'(x_0^+)$ coincide.

Corollary A. If $y = y(x)$ is a regular[3] piecewise continuously differentiable strong local minimum of the standard problem (11.1), then for some constant c,

$$L(x, y, y') - y'L_{y'}(x, y, y') = \int_a^x L_x \, dt, \quad a \leq x \leq b. \quad (11.30)$$

---

[3] A piecewise continuously differentiable extremal $y = y(x)$ is *regular* if $p = L_{y'y'}(x, y(x), y'(x)) \neq 0$ at each noncorner point x.

Proof. On each open interval I where $y = y(x)$ is continuously differentiable, y is by Hilbert twice differentiable. Hence (Exercise 11.14) there is a constant c_I so that on I,

$$L(x, y, y') - y' L_{y'}(x, y, y') = \int_a^x L_x \, dt + c_I. \tag{11.31}$$

Suppose there is a step change from left to right in the value of the constant c_I of (11.29) at the corner point x_0, i.e., $c_{I-} \neq c_{I+}$. Note that (Exercise 11.7)

$$E(x_0, y(x_0), y'(x_0^-), y'(x_0^+)) = c_{I+} - c_{I-}. \tag{11.32}$$

But (11.29) implies that
$$c_{I+} = c_{I-}. \tag{11.33}$$

The traditional proof of Result D. We deduce the second Erdmann condition from the first by means of a famous (and miraculous) change of variables. Assume $y = y(x)$ is a piecewise continuously differentiable strong local minimum of the standard problem (11.1). Choose $\alpha > 0$ sufficiently small so that

$$\alpha |y'(x)| < \frac{1}{2} \text{ on } [a,b] \text{ and } a - \alpha A < b - \alpha B. \tag{11.34}$$

Consider the linear substitution

$$x = u + \alpha v \quad \text{and} \quad y = v. \tag{11.35}$$

This substitution maps the rectangle $[a, b] \times [A, B]$ of the xy-plane bijectively onto the parallelepiped in the uv-plane shown in Figure 11.3.

Note that the resulting differential relations are (Exercise 11.8)

$$\frac{dx}{du} = 1 + \alpha v^{\#} \tag{11.36a}$$

and hence

$$y' = \frac{v^{\#}}{1 + \alpha v^{\#}}, \tag{11.36b}$$

where

$$v^{\#} = \frac{dv}{du}. \tag{11.36c}$$

11.4 Second Erdmann Corner Condition

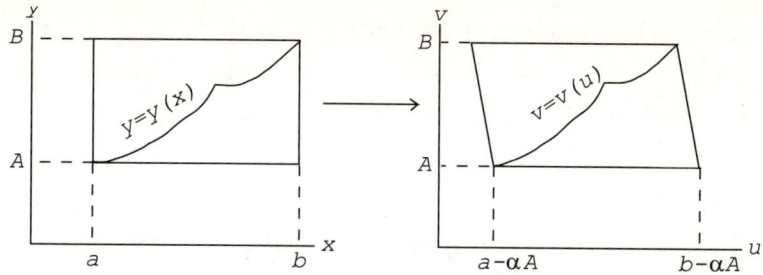

Figure 11.3 The mapping (11.35) carrys the weak local minimum $y = y(x)$ onto the weak local minimum $v = v(u)$.

The strong local minimum $y = y(x)$ is mapped onto the strong local minimum (Exercise 11.9) $v = v(u)$ on $[a - \alpha A, b - \alpha B]$ of the cost

$$J[y] = \int_{a-\alpha A}^{b-\alpha B} \Lambda(u, v, v^{\#}) \, du, \tag{11.37a}$$

subject to the end conditions

$$v(a - \alpha A) = A \quad \text{and} \quad v(b - \alpha B) = B, \tag{11.37b}$$

where the Lagrangian

$$\Lambda(u, v, w) = L(x, v, \frac{w}{1 + \alpha w}) \cdot (1 + \alpha w). \tag{11.37c}$$

Since $v = v(u)$ is an extremal of the problem (11.37), Erdmann's first corner condition guarantees that $\Lambda_{v^{\#}}$ is continuous along $v = v(u)$, that is,

$$\Lambda_{v^{\#}} = L_{y'}(x, v, \frac{v^{\#}}{1 + \alpha v^{\#}}) \cdot \frac{1}{1 + \alpha v^{\#}} + \alpha L(x, v, \frac{v^{\#}}{1 + \alpha v^{\#}})$$

$$= \frac{1 + \alpha v^{\#} - \alpha v^{\#}}{1 + \alpha v^{\#}} \cdot L_{y'}(x, v, \frac{v^{\#}}{1 + \alpha v^{\#}}) + \alpha L(x, v, \frac{v^{\#}}{1 + \alpha v^{\#}})$$

$$= L_{y'}(x, y, y') - \alpha y' L_{y'}(x, y, y') + \alpha L(x, y, y') \tag{11.38}$$

is continuous along $y = y(x)$. But the first term of the right side of (11.38) is continuous, again by Erdmann's first condition.

Corollary B. There exists a constant c so that

$$L(x, y, y') - y'L_{y'}(x, y, y') = c + \int_a^x L_x \, dt \qquad (11.39)$$

for all x in $[a, b]$.

Proof. The relation (11.39) appears in disguise (Exercise 11.15) as the first Erdmann condition

$$\Lambda_{v\#} = \int_{a-\alpha A}^{u} \Lambda_v \, dw + C. \qquad (11.40)$$

Example 3. Consider the minimization of the functional

$$J[y] = \int_{-1}^{1} y^2(1-y')^2 \, dx \qquad (11.41a)$$

subject to
$$y(-1) = 0 \quad \text{and} \quad \text{and} \quad y(1) = 1. \qquad (11.41b)$$

There can exist no continuously differentiable extremals that satisfy the end conditions (Exercise 11.10). However, the piecewise continuously differentiable

$$y(x) = \begin{cases} 0 & \text{if } -1 \leq x \leq 0 \\ x & \text{if } 0 \leq x \leq 1 \end{cases} \qquad (11.42)$$

is a global strong minimum (Exercise 11.11).

Note that both $L_{y'} = -2y^2(1-y')$ and $L - y'L_{y'}$ are zero along the extremal (11.42) and are *a fortiori* continuous.

Example 1. (revisited) The functional

$$J[y] = \frac{1}{3}\int_0^2 y'^3 \, dx \qquad (11.43)$$

has many sawtooth extremals, one type shown in Figure 11.1. Note that $L_{y'} = y'^2$ is continuous along each such extremal, which is guaranteed by the first Erdmann condition. In contrast, since none of these extremals are minima, Erdmann's second condition does not obtain, and in fact $L - y'L_{y'} = -2y'^3/3$ is not continuous.

Exercises

Example 4. (Loewen, Watson, and Wolenski, 2002) The surprisingly simple minimization of

$$J = \int_0^1 (y'^2 - 1)^2 \, dx, \qquad (11.44)$$

subject to the end conditions $y(0) = 0 = y(1)$, provides an instance where the second Erdmann corner condition (11.26) fails to hold for most of the weak local minimizers. See Exercises 11.16–11.17.

EXERCISES

11.1 Show that (11.7) implies that y is continuously differentiable.

11.2 How is Result B used in its Corollary B?

11.3 Prove the local Weierstrass condition for weak local minima (Corollary B of Result C).

11.4 Verify that the second corner condition means that the tangents to the figurative at left and right derivatives coincide.

11.5 Prove that Jacobi's necessary condition is sharp—find a weak minimum where b is conjugate to a.

Hint: Try CVP 7 on $[0, \pi]$.

11.6 Deduce the Weierstrass condition (11.22) from the Legendre condition (11.23).

Outline: Expand $E(x, y(x), y'(x), w)$ in a three-term Taylor series in w about $y'(x)$.

11.7 Verify (11.32).

11.8 Verify the differential relations (11.36).

11.9 Prove that the mapping (11.35) sends the strong local minimum $y = y(x)$ of the standard problem (11.1) onto a strong local minimum $v = v(u)$ of the problem (11.37).

11.10 Prove that there are no continuously differentiable extremals of (11.41a) that satisfy the end conditions (11.41b).

11.11 Prove that (11.42) is a global strong minimum of the problem (11.41).

11.12 Find all continuously differentiable extremals of (11.41a).

11.13 (**Project**) Restate all the results of this chapter for local maxima.

11.14 Prove (11.31). Compare with Exercise 4.53.

Outline: Using (11.2), differentiate
$$f(x) = L(x, y, y') - y' L_{y'}(x, y, y') - \int_a^x L_x \, dt.$$

11.15 Deduce (11.39) from (11.40).

11.16 Attempt to find a strong local minimum to the problem (11.44).

11.17 Identify where the traditional proof of the second Erdmann condition (Result D) fails for weak local minimizers.

11.18* A careful reexamination of the argument at the end of §4.5 will reveal that to deduce that critical curves are necessarily extremal requires a stronger du Bois-Reymond result than Exercise 4.35. Prove the following:

If for the bounded F,
$$\int_a^b F(x)\eta'(x) \, dx = 0$$
for all infinitely differentiable η with $\eta(a) = 0 = \eta(b)$, then F is constant a.e. on $[a, b]$. In fact, F equals its average value.

Outline: We may assume $a = 0$ and $b = \pi$. Then
$$0 = \int_0^\pi F(x)(\sin nx)' \, dx = n \int_0^\pi F(x) \cos nx \, dx.$$

But $1, \cos x, \cos 2x, \ldots$ form an orthogonal basis for $L^2(0, \pi)$.

11.19 Carefully prove that a piecewise continuously differentiable critical curve $y = y(x)$ of the cost (11.1) is extremal.

11.20 Prove that an n-differentiable critical curve of a continuously differentiable Lagrangian is extremal.

Appendix A
Inverse Function Theorem

The Inverse Function Theorem. Suppose

$$F(x) = (f_1(x), f_2(x), \ldots, f_n(x))^\top \tag{A.1}$$

is a vector-valued function of the vector variable $x = (x_1, x_2, \ldots, x_n)^\top$. Suppose that (each component function of) F is continuously differentiable on a neighborhood of x^0 and that its *Jacobian* at $x = x^0$ is nonzero, that is,

$$J^0 = \det\left[\frac{\partial f_i}{\partial x_j}\right]_{x^0} \neq 0. \tag{A.2}$$

Then there exists an open neighborhood U of x^0 and an open neighborhood V of $y^0 = F(x^0)$ where

$$F : U \to V \tag{A.3}$$

is bijective with a continuously differentiable inverse function $G = F^{-1}$.

Proof. We may assume $x^0 = y^0 = 0$. Introduce n new vector variables $z^j = (z_1^j, z_2^j, \ldots, z_n^j)^\top$ and set

$$J(z^1, z^2, \ldots, z^n) = \det \begin{bmatrix} \frac{\partial f_1(z^1)}{\partial z_1^1} & \frac{\partial f_1(z^1)}{\partial z_2^1} & \cdots & \frac{\partial f_1(z^1)}{\partial z_n^1} \\ \frac{\partial f_2(z^2)}{\partial z_1^2} & \frac{\partial f_2(z^2)}{\partial z_2^2} & \cdots & \frac{\partial f_2(z^2)}{\partial z_n^2} \\ & & \ddots & \\ \frac{\partial f_n(z^n)}{\partial z_1^n} & \frac{\partial f_n(z^n)}{\partial z_2^n} & \cdots & \frac{\partial f_n(z^n)}{\partial z_n^n} \end{bmatrix}. \tag{A.4}$$

Recall that continuous functions vanish on closed sets. Because $J(0, 0, \ldots, 0) = J^0 \neq 0$, there is open ball $B = \{|x| < \epsilon\}$ where

$$J(z^1, z^2, \ldots, z^n) \neq 0 \text{ for all } z^j \in B, \tag{A.5}$$

that is, J is nonzero on the cartesian product $B^n = B \times B \times \cdots \times B$. But this means that F must be injective on B, for if $F(x^1) = F(x^2)$

with $x^i \in B$, then by the mean value theorem (discussed later in this appendix),

$$0 = f_i(x^2) - f_i(x^1) = \nabla f_i(\zeta^i) \cdot (x^2 - x^1) \tag{A.6}$$

for some intermediate point ζ^i on the line segment joining x^1 and x^2. This in turn means that the vector $v = x^2 - x^1$ is in the null space of the $n \times n$ matrix

$$M = \left[\frac{\partial f_i(\zeta^i)}{\partial x_j}\right], \tag{A.7}$$

which is disallowed by (A.5) since $(\zeta^1, \zeta^2, \ldots, \zeta^n) \in B \times B \times \cdots \times B$.

Recall that continuous functions map compact sets to compact sets, and hence map closed subsets of compact sets to compact and hence closed subsets. This means an injective continuous function on a compact set necessarily possesses a continuous inverse function. Thus in our present case, when cut back to each closed subset of B, F has a continuous inverse. To see that F has a differentiable inverse requires that we cut back to a small open subset U of B as follows.

We may assume that the open ball B is so small that even its closure $\bar{B}$ is contained within an open set where F is continuously differentiable and injective. Since its boundary ∂B is compact, the image $F(\partial B)$ of this sphere has some nearest approach to 0, that is,

$$\min_{\zeta \in \partial B} |F(\zeta)| = \delta > 0. \tag{A.8}$$

Let V be the open ball about 0 of half this radius, that is,

$$V = \{y;\ |y| < \delta/2\}. \tag{A.9}$$

We now claim each point of V is taken on by F as x ranges over B. To see this, fix any point $y^* \in V$ and define

$$h(x) = |F(x) - y^*|. \tag{A.10}$$

This continuous nonnegative function h must minimize somewhere on (the compact set) $\bar{B}$, but certainly not on its boundary ∂B where $h(\zeta) > \delta/2$ since, after all, $h(0) < \delta/2$. Thus $h(x)$ and hence $h(x)^2$

The Inverse Function Theorem

must minimize at an interior point $x^* \in B$. But gradients vanish at interior extrema, giving that

$$0 = \frac{\partial h(x^*)^2}{\partial x_i} = 2 \sum_{j=1}^{n} (f_j(x^*) - y_j^*) \frac{\partial f_j(x^*)}{\partial x_i}, \quad (A.11)$$

that is, $z = F(x^*) - y^*$ is in the null space of the invertible matrix

$$N = \left[\frac{\partial f_i(x^*)}{\partial x_j} \right]^T,$$

and hence $F(x^*) = y^*$.

Let
$$U = F^{-1}(V). \quad (A.12)$$

Then we have just established that $F : U \longrightarrow V$ is a bijective bi-continuous function from the open set $U \subset B$ onto the open ball V.

Finally, the inverse function $G = F^{-1}$ is differentiable on V. To see this, let y^1 and y^0 be any two points in V and $x^i = G(y^i)$. Then by the mean value theorem as in (A.6)–(A.7),

$$y^1 - y^0 = \left[\frac{\partial f_i(\zeta^i)}{\partial x_j} \right] (x^1 - x^0),$$

and hence

$$x^1 - x^0 = \left[\frac{\partial f_i(\zeta^i)}{\partial x_j} \right]^{-1} (y^1 - y^0)$$

$$= A^{-1}(y^1 - y^0) + \left(\left[\frac{\partial f_i(\zeta^i)}{\partial x_j} \right]^{-1} - A^{-1} \right) (y^1 - y^0), \quad (A.13)$$

where
$$A = \left[\frac{\partial f_i(x^0)}{\partial x_j} \right].$$

But as $y^1 \longrightarrow y^0$, by the continuity of G, $x^1 \longrightarrow x^0$. Hence, because F is continuously differentiable, we have entry-by-entry

$$\left[\frac{\partial f_i(\zeta^i)}{\partial x_j} \right] \longrightarrow A.$$

The same limit must also hold for their inverses by the adjoint formula that gives the inverse of a matrix M as the matrix of cofactors of M divided by the determinant of M [Rade and Westergren, p. 90]. Therefore, (A.13) becomes

$$G(y) - G(y^0) = A^{-1}(y - y^0) + Z(y - y^0)$$

where
$$\lim_{y \to y^0} Z = 0.$$

Again by the adjoint formula, the entries of A^{-1}, namely $\partial G/\partial y_k$, depend continuously on y^0, giving that G is continuously differentiable on V.

The Implicit Function Theorem. Suppose

$$F(x) = (f_1(x), f_2(x), \ldots, f_n(x))^\top \qquad (A.14)$$

is a vector-valued function of the vector variable $x = (x_1, \ldots, x_{n+p})^\top$. Suppose that F is continuously differentiable on a neighborhood of x^0. Suppose further that the $n \times (n+p)$ matrix

$$M = \left[\frac{\partial f_i(x^0)}{\partial x_j}\right], \quad 1 \le i \le n,\ 1 \le j \le n+p, \qquad (A.15)$$

has full rank $r = n$. Then after a possible renumbering of the $n+p$ variables x_j and the relabeling

$$t = (t_1, t_2, \ldots, t_p)^\top = (x_{n+1}, x_{n+2}, \ldots, x_{n+p})^\top,$$

there is an open neighborhood O of $t^0 = (x^0_{n+1}, x^0_{n+2}, \ldots, x^0_{n+p})^\top$ and p continuously differentiable functions $X_i = X_i(t)$ on O such that for all $t \in O$,

$$F(X_1(t), X_2(t), \ldots, X_n(t), t_1, \ldots, t_p) = F(x^0). \qquad (A.16)$$

Proof. We may re-subscript the x_j so that the first n columns of the matrix M of (A.16) are independent. Consider the $(n+p) \times (n+p)$ vector-valued function

$$H(x) = (f_1(x), f_2(x), \ldots, f_n(x), x_{n+1}, x_{n+2}, \ldots, x_{n+p})^\top. \qquad (A.17)$$

The Inverse Function Theorem

Note that the Jacobian J_H of this map H is nonzero at $x = x^0$ since

$$J_H(x^0) = \det \begin{bmatrix} \frac{\partial f_1(x^0)}{\partial x_1} & \frac{\partial f_1(x^0)}{\partial x_2} & \cdots & \frac{\partial f_1(x^0)}{\partial x_n} & \cdots & \frac{\partial f_1(x^0)}{\partial x_{n+p}} \\ \frac{\partial f_2(x^0)}{\partial x_1} & \frac{\partial f_2(x^0)}{\partial x_2} & \cdots & \frac{\partial f_2(x^0)}{\partial x_n} & \cdots & \frac{\partial f_2(x^0)}{\partial x_{n+p}} \\ & & \ddots & & & \\ \frac{\partial f_n(x^0)}{\partial x_1} & \frac{\partial f_n(x^0)}{\partial x_2} & \cdots & \frac{\partial f_n(x^0)}{\partial x_n} & \cdots & \frac{\partial f_n(x^0)}{\partial x_{n+p}} \\ 0 & & & 1 & \cdots & 0 \\ \vdots & \cdots & & & \ddots & \vdots \\ 0 & \cdots & & & & 1 \end{bmatrix}$$

$$= \det \begin{bmatrix} \frac{\partial f_1(x^0)}{\partial x_1} & \frac{\partial f_1(x^0)}{\partial x_2} & \cdots & \frac{\partial f_1(x^0)}{\partial x_n} \\ \frac{\partial f_2(x^0)}{\partial x_1} & \frac{\partial f_2(x^0)}{\partial x_2} & \cdots & \frac{\partial f_2(x^0)}{\partial x_n} \\ & & \ddots & \\ \frac{\partial f_n(x^0)}{\partial x_1} & \frac{\partial f_n(x^0)}{\partial x_2} & \cdots & \frac{\partial f_n(x^0)}{\partial x_n} \end{bmatrix} \neq 0. \qquad (A.18)$$

Relabel the variables as follows:

$$s = (x_1, x_2, \ldots, x_n)^\top, \qquad (A.19)$$

$$t = (t_1, t_2, \ldots, t_p)^\top, \qquad (A.20)$$

$$u = (f_1(x), f_2(x), \ldots, f_n(x))^\top, \qquad (A.21)$$

with $u^0 = F(x^0)$, that is,

$$H(s^0, t^0) = (u^0, t^0)^\top. \qquad (A.22)$$

By the inverse function theorem, H possesses a continuously differentiable inverse function G mapping an open neighborhood of $(u^0, t^0)^\top$ to an open neighborhood of $(s^0, t^0)^\top$. Clearly G is of the form

$$G(u, t) = (g_1(u, t), g_2(u, t), \ldots, g_n(u, t), t_1, t_2, \ldots, t_p)^\top. \qquad (A.23)$$

Set $X_i(t) = g_i(u^0, t)$ to see that near t^0,

$$F(X, t) = F(X_1(t), X_2(t), \ldots, X_n(t), t_1, t_2, \ldots, t_p) \equiv 0.$$

The Mean Value Theorem. Suppose f is a scalar-valued function of the vector variable $x = (x_1, x_2, \ldots, x_n)^\top$ that is continuously differentiable on the open ball $B = \{x;\ |x - a| < \epsilon\}$. Then for any two points x^0 and x^1 within this ball,

$$f(x^1) - f(x^0) = \nabla f(\zeta) \cdot (x^1 - x^0) \tag{A.24}$$

for some interior point ζ of the line segment connecting x^0 and x^1.

Proof. Apply the mean value theorem in one variable and the chain rule to $g(t) = f(tx^1 + (1-t)x^0)$ to obtain

$$g(1) - g(0) = f(x^1) - f(x^0) = \frac{dg(t^*)}{dt} = \nabla f(\zeta) \cdot (x^1 - x^0),$$

where $\zeta = t^* x^1 + (1 - t^*) x^0$.

Appendix B
Picard's Theorem

Theorem. Consider the initial value problem

$$\dot{x} = F(x,t) \quad \text{subject to} \quad x(t_0) = x_0, \tag{B.1}$$

where x is a vector-valued function of t. Suppose

I. $F(x,t)$ is (jointly) continuous on an open neighborhood of (x_0, t_0), and

II. $F_x(x,t)$ is (jointly) continuous on an open neighborhood of (x_0, t_0).

Then within some open neighborhood of t_0 there exists one and only one solution $x = x(t)$ of the problem (B.1).

Proof. We may assume $t_0 = 0$ and $x_0 = 0$. To solve the initial value problem (B.1) is to find a continuous solution of the integral equation

$$x(t) = \int_0^t F(x(\tau), \tau)\, d\tau. \tag{B.2}$$

But this means we search for a function $x = x(t)$ that is invariant under the integral operator

$$(Ax)t = \int_0^t F(x(\tau), \tau)\, d\tau. \tag{B.3}$$

This integral operator A is a contraction on the complete metric space $X = C_F[-\delta, \delta]$ of all continuous functions x from $[-\delta, \delta]$ to $\mathbb{R}^n$ under the supremum norm

$$|x|_0 = \sup_{-\delta \leq t \leq \delta} |x(t)|, \tag{B.4a}$$

such that

$$|x|_0 \leq \epsilon, \tag{B.4b}$$

where δ and ϵ are chosen small enough to guarantee that all points $(x(t), t)$ lie in the neighborhoods of continuity mentioned in assumptions I and II. Such contractions have unique fixed points.

In more detail, by the mean value theorem of Appendix A,

$$|Ax - Ay|_0 = \sup_{[-\delta,\delta]} |\int_0^t F(x(\tau),\tau) - F(y(\tau),\tau)\, d\tau|$$

$$\leq \sup_{[-\delta,\delta]} |\int_0^t F_x(\xi(\tau),\tau) \cdot (x(\tau) - y(\tau))\, d\tau|$$

$$\leq \sup_{[-\delta,\delta]} \int_0^t |F_x(\xi(\tau),\tau)| \cdot |x(\tau) - y(\tau)|\, d\tau$$

$$\leq \sup_{[0,\delta]} \int_0^t b\, d\tau |x - y|_0 \leq \delta b |x - y|_0, \tag{B.5}$$

since the partials F_x are continuous and hence bounded by (say) b. Therefore, by choosing $\delta < 1/b$, we can guarantee that A is a *contraction*, that is,

$$|Ax - Ay|_0 \leq \alpha |x - y|_0 \tag{B.6}$$

for $\alpha = \delta b < 1$. Therefore, by the contraction mapping principle below, there is a unique solution to (B.1).

Contraction Mapping Principle. Suppose X is a complete metric space with distance function $d(\cdot,\cdot)$ and suppose $f : X \to X$ is a *contraction* on X, that is, for some $\alpha < 1$ and all $x, y \in X$,

$$d(f(x), f(y)) \leq \alpha\, d(x, y).$$

Then f has a unique fixed point.

Proof. Starting anywhere, at say x_0, recursively form the sequence $x_{k+1} = f(x_k)$. This sequence must be cauchy convergent since by the transitivity of the distance function $d(\cdot,\cdot)$, when $p < q$,

$$d(x_{k+q}, x_{k+p}) \leq d(x_{k+q}, x_{k+q-1}) + d(x_{k+q-1}, x_{k+q-2})$$

$$+ \cdots + d(x_{k+p+1}, x_{k+p}). \tag{B.7}$$

But

$$d(x_{m+1}, x_m) \leq \alpha d(x_m, x_{m-1}) \leq \cdots \leq \alpha^m d(x_1, x_0). \tag{B.8}$$

Picard's Theorem

Thus
$$d(x_{k+q}, x_{k+p}) \le d(x_1, x_0) \sum_{i=k+p}^{k+q-1} \alpha^i, \tag{B.9}$$

and thus cauchy convergence is inherited from the convergence of the geometric series.

Let x_∞ denote the limit of the x_k. Then x_∞ is a fixed point of f since, because of the sequential continuity of f,
$$x_\infty = \lim_{k\to\infty} x_{k+1} = \lim_{k\to\infty} f(x_k) = f(\lim_{k\to\infty} x_k) = f(x_\infty).$$

There cannot be two distinct fixed points y_1 and y_2 of f since
$$d(y_1, y_2) = d(f(y_1), f(y_2)) \le \alpha d(y_1, y_2).$$

Remark. In Picard's theorem we may relax the hypotheses I and II to half intervals of the independent variable. That is, we need assume only that F and F_x are continuous on sets of the form
$$\{|x - x_0|_0 < \epsilon\} \times [t_0, t_0 + \delta).$$

Modify the proof by replacing the closed intervals $[-\delta, \delta]$ by $[0, \delta]$. The result now guarantees a unique solution on some interval $[t_0, t_0 + \delta)$ with a right derivative at $t = t_0$.

Corollary. The problem
$$\dot{x} = F(x, u(t)) \quad \text{subject to} \quad x(t_0) = x_0, \tag{B.10}$$

where F and F_x are both continuous in a neighborhood of $(x_0, u(t_0))$, has a unique (continuously differentiable) solution $x = x(t)$ near $t = t_0$, provided u is continuous on an open neighborhood of t_0.

Proof. Let $G(x(t), t) = F(x(t), u(t))$. Apply Picard's theorem to G.

Question. If a differential equation and its initial condition are slightly perturbed, will the solutions remain nearby?

The Perturbation Lemma. Suppose x and y are solutions on $[a, b]$ to the respective problems
$$\dot{x} = F(x, t) \text{ subject to } x(a) = x_0, \tag{B.11a}$$
$$\dot{y} = G(y, t) \text{ subject to } y(a) = y_0. \tag{B.11b}$$

Assume both F and G are jointly continuous on some open set containing the (compact) set $C \times [a, b]$, where C is some compact set containing the paths $x = x(t)$ and $y = y(t)$ on $[a, b]$. Suppose further that F satisfies the Lipschitz condition

$$|F(x_2, t) - F(x_1, t)| \leq \kappa |x_2 - x_1| \tag{B.12}$$

on $C \times [a, b]$. Then on $[a, b]$,

$$|y(t) - x(t)| \leq K\left(1 + (t - a)\kappa e^{\kappa(t-a)}\right), \tag{B.13a}$$

where

$$K = |y_0 - x_0| + (b - a) \sup_{C \times [a,b]} |G(z, t) - F(z, t)|. \tag{B.13b}$$

Proof. By rewriting the differential equations (B.11) as integral equations and subtracting, we have

$$|y(t) - x(t)| = |y_0 - x_0 + \int_a^t G(y(\tau), \tau) - F(x(\tau), \tau)\, d\tau|$$

$$\leq |y_0 - x_0| + \int_a^t |G(y(\tau), \tau) - F(y(\tau), \tau)|\, d\tau + \int_a^t |F(y(\tau), \tau) - F(x(\tau), \tau)|\, d\tau$$

$$\leq K + \int_a^t |F(y(\tau), \tau) - F(x(\tau), \tau)|\, d\tau \leq K + \kappa \int_a^t |y(\tau) - x(\tau)|\, d\tau. \tag{B.14}$$

Employ the **"E-trick"**: Let

$$E(t) = \int_a^t |y(\tau) - x(\tau)|\, d\tau. \tag{B.15}$$

Then our estimate (B.14) becomes simply

$$\frac{d}{dt}E(t) \leq K + \kappa E(t). \tag{B.16}$$

Multiplying through by the integrating factor $\mu = e^{-\kappa(t-a)}$ yields

$$\frac{d}{dt}\left(e^{-\kappa(t-a)} E(t)\right) \leq K e^{-\kappa(t-a)} \leq K, \tag{B.17}$$

which integrates to

$$E(t) \leq K(t - a)e^{\kappa(t-a)}. \tag{B.18}$$

Placing (B.18) into (B.16) yields our result.

Thus if both the dynamical rules F and G and the initial values are close, so are the solutions. The estimate (B.13) can be tightened at many junctures in the proof.

Appendix C
The Divergence Theorem

Theorem. Suppose F is a continuously differentiable vector field on a domain containing the closed orientable surface S that encloses the volume V. Then the flux of F leaving S is the integral over V of the divergence of F, that is,

$$\int_S F \cdot \mathbf{n} \, d\sigma = \int_V \nabla \cdot F \, dV.$$

Proof sketch. Partition V in the usual way into small congruent cubes by equally spaced planes parallel to the coordinate planes. For all interior cubes, the flux leaving each face enters the contiguous cube and so summing over all cubes C that meet V,

$$\sum_C \int_{\partial C} F \cdot \mathbf{n} \, d\sigma = \sum_{C'} \int_{\partial C' \cap S} F \cdot \mathbf{n} \, d\sigma \approx \int_S F \cdot \mathbf{n} \, d\sigma,$$

where the second sum is over all faces of the cubes C' that meet the boundary S. Moreover,

$$\sum_C \int_C \nabla \cdot F \, dV = \sum_C \int_C \nabla \cdot F \, dV \approx \int_V \nabla \cdot F \, dV,$$

where in both cases the approximation becomes exact as the partition into cubes is refined. So it is sufficient to prove the theorem for the cube $C : a \leq x \leq b, \ c \leq y \leq d, \ e \leq z \leq f$. But if $F = P\mathbf{i} + Q\mathbf{j} + R\mathbf{k}$, by the fundamental theorem of calculus,

$$\int_C \nabla \cdot F \, dC = \int_a^b \int_c^d \int_e^f P_x + Q_y + R_z \, dz \, dy \, dx$$

$$= \int_c^d \int_e^f P(b, y, z) - P(a, y, z) \, dz \, dy + \cdots$$

$$= \int_{x=b} F \cdot \mathbf{i} \, d\sigma + \int_{x=a} F \cdot \mathbf{i} \, d\sigma + \cdots = \int_{\partial C} F \cdot \mathbf{n} \, d\sigma.$$

Appendix D
A MATLAB Cookbook

What follows is a crash course in the exceptional package MATLAB from [MathWorks].

D.0 Interactive versus Scripts

There are two ways to interact with MATLAB: *interactively* and via *scripts*. At the MATLAB prompt, typing a request followed by ENTER will return the execution of the request. In the next line, results from previous lines can be referenced and manipulated. This is step-by-step interaction with MATLAB.

Alternatively, when the computation is composed of a series of many calculations or will be repeated and refined, it is better to prepare a *script*—where all steps of the routine are prepared in advance with a plaintext editor like *Notepad*, with the script saved as a **.m** file, such as **exper.m**. Typing the script file name at the MATLAB prompt and hitting ENTER will cause MATLAB to execute the steps of the script in order. Such editors are included with most implementations of MATLAB.

D.1 Writing Arrays

The line of code v = [1,2,3,4,5] assigns the symbol v to a row vector with entries 1,2,3,4,5. This is also accomplished simply by setting v = 1:5. Nonintegral steps can be done with v = 1:0.1:5 to accomplish v = [1, 1.1, 1.2,...,4.9, 5]. Alternatively, v = linspace(x1,x2,n) returns with a row vector of length n of equally spaced values between $x1$ and $x2$. So for example, 1:0.1:5 = linspace(1,5,41).

Matrices are assigned row by row, separated by semicolons. Row entries are separated by commas, or optionally, a space. For example, the 3×3 identity matrix can be assigned to the symbol I with

I = [1,0,0;0,1,0;0,0,1], or I = [1 0 0; 0 1 0; 0 0 1],

or by using the built-in command I = eye(3). Or for another example, M = [1 3 1; 4 5 6] assigns M to the 2×3 matrix

$$\begin{bmatrix} 1 & 3 & 1 \\ 4 & 5 & 6 \end{bmatrix}.$$

There *many* such built-in commands for generating special matrices, for example M = diag(v) and v = diag(M).

A linear system of n equations in n unknowns, when written in matrix form $Ax = b$, is solved with the single request x = A\b via Gaussian elimination [which is more advisable than requesting x = inv(A)*b]. This same command x = A\b will yield solutions in the least-square's sense for any overdetermined or underdetermined systems $Ax = b$.

Eigenvalues of a (square) matrix A are obtained with d = eig(A), which returns a row vector d of the eigenvalues of A. There are many built-in useful extensions of the command eig.

Matrices A, B of congruent shapes are added with the request A + B. Scalar muliplication is achieved with c*A. Matrix multiplication of an $m \times n$ matrix A followed by a $n \times p$ matrix B is accomplished with C = A*B to obtain the $m \times p$ matrix C. Taking the transpose of A is accomplished with a prime A'.

If A and B are of congruent shape, then A.*B returns the matrix of elementwise multiplication. In general, the dot before an operation yields the elementwise operation. For instance, u.^v returns the vector where each entry of u has been raised to the power given by the corresponding entry of v. If a request is mathematically nonsensical, then MATLAB often will return with an error message. In other cases, it will execute the most reasonable interpretation; for example, the request v - 1, where v is a vector, will return with a vector where 1 has been subtracted from each component of v.

D.2 Scalar versus Vector Computations

As a rule, "do loops" are replaced in MATLAB by vector calculations.

A MATLAB Cookbook

Task A: Computing a List of Squares

```
% scalar                    % vector
N=10;                       x = 1:10;
for i=1:N                   y = x.*x
   y(i) = i*i;
   end;
y
```

Task B: Graphing One Period of y = sin x

```
% scalar                    % vector
N = 100;                    x = linspace(0,2*pi,101);
dx = 2*pi/N;                y = sin(x);
for i = 1:N                 plot(x,y)
   x(i) = i*dx;
   y(i) = sin(x(i))
   end;
plot(x,y)
```

In general, the plot command plot(u,v) will plot the row (or column) of v as ordinate against the row (or column) of u as abscissa. If, say, v is a matrix, multiple superimposed plots will result. For a matrix M, plot(M) will superimpose plots of the columns of M against an abscissa labeled by row numbers.

Task C: Enhancing the Graph of y = sin x

Two graphs may be superimposed with the hold command. For example, to superimpose the graph of $y = J_1(x)$ onto the graph of $y = \sin x$ obtained in Task B, merely add the lines

```
% superimposing graphs
hold on;
J = bessel(1,x);
plot(x,J)
```

Rather than plotting points connected by straight-line segments (default), we may plot instead discrete points with plot(x,y,.), or dashed lines with plot(x,y,--), and so forth; many styles are available. We may also choose different colors for each superimposed graph.

We may title a graph and its axes with say title('The graph of y = sin(x)'), xlabel('x'), and ylabel('y') as well as place text within the graph at specified locations with text. MATLAB provides many tools for contructing superior presentation graphics. We can add to the list of standard functions by storing a *function script* (see §D.0).

Task D: Graphing a Parametric Curve

Graph the 3-D helix $x = \cos t$, $y = \sin t$, $z = t$ with the script

```
% parametric plots
t = linspace(0,6*pi);
x = cos(t);
y = sin(t);
z = t;
plot3(x,y,z)
```

Task E: Graphing a 3-D Surface

Graph the sinc function $z = (\sin r)/r$ (in cylindrical coordinates) with the script

```
% mesh plot
a = 3*pi;
x = linspace(-a,a);
y = x;
[X,Y] = meshgrid(x,y);
R = sqrt(X.*X + Y.*Y) + 0.0001;
Z = sin(R)./R;
mesh(Z)
```

There are many built-in means to massage these plots.

D.3 Monte Carlo Experiments

Rather than obtaining an analytic solution to a problem, it is often enough merely to instruct a computer to perform experiments to build up a statistical picture of the phenomenon. The traditional first example is "dartboard integration."

A MATLAB Cookbook

Task A: Integration

Let us estimate the integral

$$\int_0^1 e^{-x^3} dx$$

by choosing random points within the rectangle $0 \le x, y \le 1$ enclosing the curve $y = \exp(-x^3)$. The number of hits under the curve divided by the area of the rectangle must approximate the area under the curve:

```
% Monte Carlo integration
A = 1;                      % area of enclosing rectangle
N = 10000;                  % number of experiments
s = 0;                      % initialize success counter
for i=1:N                   % do N experiments
  x = rand;                 % pick a random x coordinate
  y = rand;                 % pick a random y coordinate
  if y <= exp(-x^3);        % if .., then
    s = s + 1;              % increment success counter
    end;                    % end if
  end;                      % end each experiment
I = s*A/N                   % print approximate value
```

The routine will return an estimate of $I \approx 0.805$. The command `rand` returns a random number uniformly distributed between 0 and 1.

There are of course superior numerical methods of integration for functions of one variable. Monte Carlo methods become more practical for integrating functions of many variables.

Task B: Mean Time between Failures

A manufactured product is composed of three components, each with a normally distributed time before failure of mean $\mu = 2$ years and standard deviation $\sigma = 0.75$. If the construct fails whenever any one of its three components fail, what is the expected life of this product?

We merely build the product many times and average the failure times:

```
% MTBF
N = 10000;                  % number built
m = 0;                      % initialize failure time
for i=1:N                   % do N experiments
  x = 2 + 0.75*randn;       % random lifetime of 1st component
  y = 2 + 0.75*randn;       % random lifetime of 2nd component
  z = 2 + 0.75*randn;       % random lifetime of 3rd component
  w = min([x,y,z]);         % failure at minimum lifetime
  m = m + w/N;              % add in life for this experiment
  end;                      % end this experiment and loop back
m                           % print mean lifetime
```

The command **randn** returns with a normally distributed random number with mean $\mu = 0$ and standard deviation $\sigma = 1$.

Task C: Testing Distributions

Let us test MATLAB's Gaussian distribution generator by gathering its output into bins.

```
% bins of data
N = 10000;                  % number of random calls
b = -4:0.1:4;               % bins defined
x = randn(1,N);             % a random N row vector
hist(x,b)                   % histogram of bin hits
```

D.4 Images

Every $m \times n$ matrix $X = [x_{ij}]$ with entries $0 \leq x_{ij} \leq W$ can be thought of as a black-and-white image, consisting of mn pixels, each of grayscale intensity x_{ij}, where 0 is black, W is white.

Task A: Display a Matrix as an Image

The Toeplitz matrix formed from the vector $r = [1, 2, 3, \ldots, 100]$ can be visualized as an image with the following script:

```
% matrix as an image
r = 1:100;
T = toeplitz(r);
image(T)
axis off
colormap gray
```

A Matlab *Cookbook*

Task B: Importing a Photo

Consider the digital photo of two llamas, Zack and Bandit, given as a matrix of pixels, stored as the file **llama.jpeg**. At each pixel is a triple of red, green, blue intensities that form the color photo. This photo can be downloaded from

http://www.math.msu.edu/~maccluer/Farm/llama.jpeg

```
X = imread('llama.jpeg');
image(X)                    % the color photo will appear
axis off
```

This photo can be converted to grayscale with the additional steps:
```
% converting to grayscale
X = double(X);
Y = X(:,:,1)+X(:,:,2);
Y = Y + X(:,:,3);
X = Y/3;                    % adjust brightness
image(X)
axis off
colormap gray
```

Task C: Enhancing Contrast

Suppose X is a $m \times n$ matrix of grayscale pixels. We can modify the contrast of this image by various schemes, for instance, by *thresholding*, where each pixel is set to either black or white, depending on its orginal intensity.

```
% thresholding
m = max(max(X));
Z = (X > m/2);              % returns 1 if T, 0 if F, entry-by-entry
image(m*Z)
axis off
colormap gray
```

Task D: Filtering Images

An image X is filtered by convolving the image with a "deblurring" matrix H, producing the hopefully improved image $Y = H * X$. This convolution is best performed by first transforming to the frequency domain by means of the *discrete Fourier transform* (DFT), implemented via the *fast Fourier transform* (FFT). Once transformed, convolution becomes elementwise multiplication. Finish with the inverse DFT. See [MacCluer, 2000].

```
% deblurring grayimage X with filter H
Xhat = fft2(X);          % the DFT of X
Hhat = fft2(H);          % the DFT of H
Yhat = Hhat.*Xhat;       % entry-by-entry multiplication
Y = ifft2(Yhat);         % inverse DFT
Y = real(Y);             % clean away small imaginary parts
image(Y);                % display the deblurred image
axis off
colormap gray
```

Task E: Morphing One Image into Another

We may slowly "morph" one image onto another via a homotopy.

```
% morphing X onto Y
fps = 3;                      % set frames per second
N = 100;                      % number of frames
for k = 0:N
  Z = (1-k/N)*X + (k/N)*Y;    % the homotopy
  image(Z);                   % compute image
  axis off
  M = getframe;               % store this frame
  end;                        % loop back for next frame
                              % play movie M once at
movie(M,1,fps)                % fps frames per second
```

D.5 Sounds

MATLAB provides the capability to 'hear' a vector. The entries of the vector are interpreted as sound pressure measurements sampled at equally spaced intervals of time.

Task A: Produce a Single Tone

```
% pure tone
fS = 8192;                % sampling freq Hz
L = 5;                    % length of tone in seconds
t = 0:L*fS-1;             % sampling vector size for L sec
t = t/fS;                 % sampling times
w = 2*pi*440;             % radian freq for 440 Hz
x = sin(w*t);             % a pure tone
wavwrite(x,fS,'tone');    % create the wave file tone.wav
```

This wave file **tone.wav** can now be played by the soundcard.

A MATLAB Cookbook   243

Task B: Produce a Soothing Sound

```
% random noise
fS = 8192;                  % sampling freq Hz
L = 5;                      % length of tone in seconds
n = randn(1,L*fS);          % random row vector
n = 0.2*n;                  % reduce volume
wavwrite(n,fS,'noise');     % create a wave file noise.wav
```

Task C: Synthesize an Oboe

The idea is to add overtones and wind noise to the fundamental tone to approximate the spectral content, and hence the sound of the actual instrument.

```
% oboe
fS = 8192;                       % sampling freq Hz
L = 5;                           % length of tone in seconds
t = 0:L*fS-1;                    % sampling vector for L s
t = t/fS;                        % sampling times
w = 2*pi*440;                    % radian freq for 440 Hz
x = sin(w*t)+ 0.1*sin(2*w*t);    % fund. plus 2nd harmonic
x = x + 0.1*sin(2.8*w*t);        % flat 3rd
x = x + 0.1*sin(4.2*w*t);        % sharp 4th
x = x + 0.4*randn(1,L*fS);       % wind noise
wavwrite(x,fS,'oboe');           % create the file oboe.wav
```

As an exercise you are asked to refine this synthesis, choosing additional overtones and adjusting amplitudes, until the sound is lifelike.

D.6 Spectral Analysis

The spectrum of data often reveals clues about the data. For example, the price history of a commodity may be composed of hidden superimposed cycles and hence be price predictable. The spectrum of sampled data is obtained via the discrete Fourier transform (DFT), implemented as the fast Fourier transform (FFT) [MacCluer, 2000].

Task A: Find the Spectrum of Recorded Data

Assume we have sampled data taken at N equal time intervals T given as the row vector x.

```
% spectral analysis
xhat = fft(x);              % perform the DFT
r = abs(xhat)/N;            % modulus of spectrum
plot(r)                     % display spectrum abs value
```

The preceeding rather crude spectral display should be refined:

```
% refined display
fS = 1/T;                   % compute sampling frequency
f = linspace(0,fS,N);       % abscissa in hertz
m = 400;                    % min viewing freq (Hz)
mf = round(m*N/fS);         % rescale
M = 2200;                   % max viewing freq (Hz)
Mf = round(M*N/fS);         % rescale
plot(f(mf:Mf),r(mf:Mf))
```

You will notice that the spectrum magnitude is symmetric about half the sampling frequency, an artifact of sampling itself—*It is impossible to distinguish frequencies that are congruent modulo the sampling frequency.* See [MacCluer, 2000].

D.7 Defining Functions

A user can add to MATLAB's zoo of build-in functions by storing a *function script* in the working directory.

Task A: Graph y = sinc(x)

Store on disk a file **sinc.m** consisting of the lines

```
function y = sinc(x)
  y = sin(x)./x;
```

We may now call the function sinc(x) in subsequent calculations, as in

```
% using fplot
fplot('sinc',[-5*pi,5*pi])
```

or

```
% integration
quad('sinc',eps,5*pi)
```

A MATLAB Cookbook

D.8 Input/Output

To save all the variables constructed during a session, use the `save` command. For example, `save session1` will save all variables to the disk as the MATLAB formatted file **session1.mat**. All these stored values can be imported during a later session with the command `load session1`.

Most useful is the ability to import tables of ASCII data generated by other software tools. The command `load file.ext` followed by `A = file` will bring a table of ASCII data stored on disk as the file (say) **file.ext** into a MATLAB session as the matrix `A`.

All subsequent steps during a session can be recorded for later perusal with the command `diary`.

D.9 Obtaining Help

Online help is available at the MATLAB prompt, for example, `help plot` returns with the reference manual page on the `plot` command. The *MATLAB Reference Guide* is obtainable from Mathworks, Inc. Mathworks online help is available at
http://www.mathworks.com/access/helpdesk/help/helpdesk.shtml

See the excellent references by [Loan], [Lindfield and Penny], and [Knight].

EXERCISES

D.1 Using MATLAB, solve (if possible) the system

$$\begin{bmatrix} 1 & 2 & -1 & 3 & -2 \\ -1 & 3 & 2 & 6 & 2 \\ 1 & 1 & 6 & 4 & 3 \\ 2 & -2 & 2 & 5 & -1 \\ 7 & 0 & 5 & 3 & 3 \end{bmatrix} \begin{bmatrix} x_1 \\ x_2 \\ x_3 \\ x_4 \\ x_5 \end{bmatrix} = \begin{bmatrix} 1 \\ -3 \\ 4 \\ -7 \\ 2 \end{bmatrix}.$$

Check your answer. Experiment and report on solving overdetermined and underdetermined systems.

D.2 Find the eigenvalues and eigenvectors of the above matrix.

D.3 Consider the $n \times n$ array $H = [1/(i+j-1)]$, the infamous ill-conditioned *Hilbert matrix*. Find its condition number (via `cond`) and inverse (via `inv`) for $n = 1, 2, \ldots$, for as long as MATLAB can hold off the numerical instabilities. Is computing `H\I` a better approach?

D.4 Using `norm`, compute the L^2 norm of the 5×5 matrix of Exercise D.1. Find a unit vector v that realizes this norm, that is, where $\|Av\| = \|A\|$.

D.5 Superimpose the graphs of the first eight Bessel functions $J_0, J_1, \ldots, J_7$ for $0 \leq x \leq 6\pi$. What do you notice about their zeros?

D.6 Write your own script to find all primes less than N. Do not use `isprime`.

D.7 Using `rand`, prepare a list of 100 random integers between 1 and 1000. Sort this list into ascending order using `sort`.

D.8 Using `roots`, find the zeros of $p(x) = 5x^4 - 3x^2 + 2x - 7$. Perversely, find these zeros by instead using `eig`.

D.9 Prepare a carefully labeled 2-D graph for classroom use to communicate an important concept.

D.10 Vectorize the routine of Task B in §D.3.

D.11 Superimpose the graphs of $y = \operatorname{sgn} x$ and the first 11 terms of its Fourier series on $[-\pi, \pi]$. Observe the Gibbs phenomenon at the discontinuities.

D.12 Numerically approximate via `quad` the value of the elliptic integral
$$I = \int_0^{\pi/2} \frac{d\theta}{\sqrt{4 - \sin^2 \theta}}$$
to 5 places. Compare with your own Simpson routine.

D.13 Numerically solve via `ode45` and plot selected solution trajectories $(x(t), y(t))$ to the system of ODEs
$$\dot{x} = x + y - x(x^2 + y^2)$$
$$\dot{y} = -x + y - y(x^2 + y^2).$$

D.14 Experimentally verify that with probability 1, all square matrices are invertible.

D.15 Simulate the sound of a siren.

D.16 Assemble numerical evidence that the harmonic series diverges.

D.17 Experiment with Task C of §D.5 to synthesize more accurately the sound of an oboe or another instrument of your choice.

D.18 Vectorize Task E of §D.4.

D.19 Intentionally blur an image X by convolving with a random blurring matrix B. Display the encrypted image $Y = B * X$. Next deblur and display. Has the image lost definition?

D.20 Prepare a contour map of a surface using `contour3`.

D.21 Prepare a detailed anotated classroom slide of a surface that illustrates an important fact.

D.22 Is Gaussian noise more soothing than uniformly distributed noise? What is your evidence?

D.23 Make a movie that slowly morphs one animal photo into another.

D.24 Write a script that finds the greatest common divisor of any two integers via the Euclidean algorithm.

A MATLAB Cookbook

D.25 Fit the data

x	1.1	1.3	1.5	1.7	1.9
y	3.4	4.4	6.7	9.2	7.3

with a cubic spline via `spline` (see [MacCluer, 2000]). Graph the results with the knots displayed.

D.26 Fit the data

x	1.1	1.3	1.5	1.7	1.9	2.1
y	3.4	4.4	6.7	9.2	7.2	5.1

with a quintic polynomial using polynomial regression via `polyfit`. Graph the fit. Compare with the fit of a spline interpolation via `spline`.

D.27 Using the the cubic spline routine `spline`, repair $f(x) = |x|$ so that it becomes differentiable on $(-1, 1)$.

D.28 Write your own script to solve $y' = x \sin y^2$ subject to $y(0) = \sqrt{\pi/2}$ on for $0 \leq x \leq 1$ via the Euler, improved Euler (Heun), and Runge-Kutta numerical methods. Tabulate your results. Compare with results from `ode45`.

D.29 Make a movie of a wave traveling up and down a clothesline.

D.30 The DFT $\hat{X}$ represents the frequency content of the image X, where the lower-frequency components of X are found in the upper-left-hand corner of $\hat{X}$ (and in the aliased congruent lower-right-hand corner). Lowpass filter a grayscale image X to form an image Y by setting all entries to 0 that lie outside an upper left hand block of $\hat{X}$ to form $\hat{Y}$. Compare its inverse transform Y to the original image X. What aspects of the original image are lost in this filtering?

In contrast, normalize each (complex) entry of $\hat{X}$ to modulus 1 and consider the inverse of the result. What aspects of X are preserved?

D.31 Revisit Exercise D.23. A matrix A is choosen at random by a real-world computer of finite precision. What is the exact probability that A is invertible?

D.32 (**Project**) Attempt to best MATLAB's equation solver. Using large (ill-conditioned) Hilbert matrices $H = (1/(i+j-1))$, select a vector x_0 and set $b = Hx_0$. Write a routine in C to solve any system $Ax = b$ for x, apply this routine to $A = H$, and compare with MATLAB's answer `x = H\b`. Compare your accuracy against the *truth model* x_0 with various norms, for example, largest error in any one coordinate ($\|x - x_0\|_\infty$), the sum of absolute errors over all components ($\|x - x_0\|_1$), or especially the square root of the sum of the squares of component errors ($\|x - x_0\|_2$). Can you achieve even 20% of MATLAB's accuracy?

D.33 (**Project**) Make a stick-figure movie. Your result will be judged on artistic as well as technical merit. Add sound.

D.34 (**Project**) Solve the *three-body problem:* Model, code, and simulate the motions of a sun and two massive interacting planets. Do the special case where the sun is essentially stationary and motion is planar. Assume all three masses to be point masses. Will collisions occur? Will a planet spiral in to be consumed by the sun? Can a planet escape the system? Plot selected motions. Make a movie of several of the more interesting motions.

D.35 (**Project**) Using MATLAB, make a movie of the vibrations of the square drum governed by the wave equation

$$u_{tt} = \nabla^2 u, \qquad 0 < x, y < 1$$

subject to zero boundary conditions. Play the sound of this drum. See [MacCluer, 2004] for a *Mathematica*-generated movie.

D.36 (**Project**) (G. C. Stockman) What is the least number of pixels and grayscale levels needed to recognize familiar faces? Take an image of Abraham Lincoln or Marilyn Monroe from the Web. Reduce the number of pixels and greyscale level until the image is just recognizable, keeping the size of the image constant. It helps to blur the sharp edges as pixels become coarser. Recognition seems possible down to 32×32 pixels and 3-bit grayscale. For example, if we start with a 256×256 image of 256 gray levels, then we can next try a 128×128 image of 256 gray levels formed as follows: Each 2×2 block of the input image is averaged and the average is repeated 4 times in the output image—the output still has 256×256 pixels, but each is really a 2×2 block of identical gray levels.

D.37 (**Project**) Analyze the two- or three-year price history of a commodity for hidden cycles after first factoring out (exponential) inflation and price creep.

D.38 Contrary to popular belief, the period T of oscillation of the planar pendulum (CVP 8) is not independent of its maximal angle θ_0 of excursion. Larger θ_0 yield larger periods T; in the extreme case, $T = \infty$ when $\theta_0 = \pi$. In fact,

$$\frac{T}{4} = \sqrt{\frac{a}{2g}} \int_0^{\theta_0} \frac{d\theta}{\sqrt{\cos\theta - \cos\theta_0}} = k\sqrt{\frac{a}{g}} \int_0^{\theta_0/2} \frac{d\phi}{\sqrt{1 - k^2 \sin^2\phi}},$$

where $k = \sin(\theta_0/2)$.

Provide numerical evidence that in the limit as $\theta_0 \to 0$, the period T approaches the period $2\pi\sqrt{a/g}$ of the harmonic oscillator $\ddot\theta = -(g/a)\theta$, a model close to the pendulum

$$\ddot\theta = -(g/a)\sin\theta$$

for small excursions θ.

Prepare a table of pendulum periods T against θ_0.

References

T. M. Apostol, *Mathematical Analysis*, Addison-Wesley, Reading, MA, 1960.

M. Athens and P. L. Falb, *Optimal Control*, McGraw-Hill, New York, 1966.

D. J. Bell and D. H. Jacobson, *Singular Optimal Control Problems*, Academic Press, New York, 1975.

V. G. Boltyanskii, R. V. Gamkrelidze, and L. S. Pontryagin (1956), *On the Theory of Optimal Processes*, Doklady Akad. Nauk SSSR (in Narhoski et al., 1983).

V. G. Boltyanskii (1956), *The Maximum Principle in the Theory of Optimal Processes*, Doklady Akad. Nauk SSSR (in Nahorski et al., 1983).

Wm. C. Brown, *A Second Course in Linear Algebra*, J. Wiley and Sons, New York, 1988.

B. Y. Chen, *Riemannian Submanifolds, Handbook of Differential Geometry*, vol. I, North Holland Publ., 2000, pp. 187–418.

F. H. Clarke, *Calculus of Variations and Optimal Control*, unpublished lecture notes, Univ. British Columbia, 1979.

R. P. Feynman, *Statistical Mechanics*, Addison-Wesley, Redwood City, CA, 1972.

W. Fleming and R. Rishel, *Deterministic and Stochastic Optimal Control*, Springer-Verlag, Berlin, 1975.

G. B. Folland, *Advanced Calculus*, Prentice Hall, Upper Saddle River, NJ, 2002.

C. Fox, *The Calculus of Variations*, Dover Publications, New York, 1987.

I. M. Gelfand and S. V. Fomin, *Calculus of Variations*, Dover, NY, 2000.

S. H. Gould, *Variational Methods for Eigenvalue Problems*, Dover, NY, 1995.

V. J. Katz, *History of Mathematics; an introduction*, Harper Collins, New York, 1993.

M. Kline, *Mathematical Thought from Ancient to Modern Times*, Oxford University Press, New York, 1972.

A. Knight, *Basics of MATLAB and Beyond*, Chapman and Hall, CRC, Boca Raton, FL, 2000.

G. Knowles, *An Introduction to Applied Optimal Control*, Academic Press, New York, 1981.

S. G. Krantz, *Real Analysis and Foundations*, CRC Press, Boca Raton, FL, 1991.

H. Kwakernaak and R. Sivan, *Linear Optimal Control Systems*, Wiley Interscience Publishers.

G. Lindfield and J. Penny, *Numerical Methods using MATLAB*, 2nd ed., Prentice Hall, Upper Saddle River, NJ, 2000.

C. F. Loan, *Introduction to Scientific Computing*, 2nd ed., Prentice Hall, Upper Saddle River, NJ, 2000.

P. D. Loewen, S. J. Watson, and P. R. Wolenski, "A variational problem with a continuum of weak local minima," *Proceedings of the 41st IEEE Conference on Decision and Control*, Las Vegas, 2002, pp. 10–13.

C. R. MacCluer, "On Extreme Points of the Numerical Range of Normal Operators," *Proc. AMS,* Vol. 16, No. 6, December 1965, pp. 1183–1184.

C. R. MacCluer, *Industrial Mathematics,* Prentice Hall, Upper Saddle River, NJ, 2000.

C. R. MacCluer, *Boundary Value Problems and Fourier Expansions,* Dover, NY, 2004.

J. Macki and A. Strauss, *Introduction to Optimal Control Theory,* Springer-Verlag, New York, 1982.

Mathworks Inc., *MATLAB Reference Guide,* Mathworks, Natick, MA, 2000.

E. J. McShane, "The Calculus of Variations from the beginning through Optimal Control Theory," *SIAM J. on Control and Optimization,* **27** (5) (1989), pp. 916–939.

J. A. Mirrlees, "An Exploration in the Theory of Optimum Income Taxation," *The Review of Economic Studies,* Vol. 38, Issue 2 (April 1971), 175-208.

J. Oprea, *The Mathematics of Soap Films: Explorations with Maple,* AMS Student Mathematical Library **10**, Providence, RI, 2000.

R. Osserman, *A Survey of Minimal Surfaces,* Dover, New York, 1986.

D. A. Pierre, *Optimization Theory with Applications,* Dover Publications, New York, 1969.

L. Pontryagin, V. G. Boltyanskii, R. V. Gamrelidze, and E. F. Mishchenko, *The Mathematical Theory of Optimal Processes,* Oxford, New York, Pergamon Press; [distributed in the Western Hemisphere by Macmillan, New York], 1964.

L. Rade and B. Westergren, *Beta Mathematics Handbook,* 2nd. ed., CRC Press, Boca Raton, FL, 1990.

J. W. S. Rayleigh, *The Theory of Sound,* 2nd.ed., vol. 1, Dover, New York, 1945.

J. N. Reddy, *An Introduction to the Finite Element Method,* 2nd ed., McGraw-Hill, Boston, 1993.

M. Reed and B. Simon, *Functional Analysis,* Methods of Modern Mathematical Physics I., Academic Press, San Diego, 1980.

W. Rudin, *Real and Complex Analysis,* 3rd ed., McGraw-Hill, New York, 1966.

H. Sagan, *Boundary and Eigenvalue Problems in Mathematical Physics,* Dover Publications, New York, 1989.

H. M. Schey, *Div, Grad, Curl, and All That,* W. W. Norton, New York, 1973.

G. Strang and G. Fix, *An Analysis of the Finite Element Method,* Prentice Hall, Upper Saddle River, NJ, 1973. B. L. van der Waerden, *Sources of Quantum Mechanics,* Dover, New York, 1968.

W. R. Wade, *An Introduction to Analysis,* 2nd ed., Prentice Hall, Upper Saddle River, NJ, 2000.

R. Weinstock, *Calculus of Variations,* Dover, New York, 1974.

O. C. Zienkiewicz, *The Finite Element Method in Engineering Science*, McGraw-Hill, New York, 1971.

Index

A

a.b.f.n.
  all but a finite number, 121
Absolutely continuous, 59
Accessory equation, 171, ex. 9.30[1]
Action integral, ex. 3.22
Admissible control, 112
a.e., almost every
Affine, 123
Assymptotically stable, 143
Average voltage (OCP 3), 115
Arrays, 235

B

Bang-bang,
  control, 116
  theorem, 118
Bauer's maximum principle,
  ex. 7.29
Benford's Law, ex. 2.33
Boundary condition,
  end, ex. 4.36
  natural, ex. 4.36
Brachistochrone (CVP 4), 33
Bushaw's example, 127

C

Cam, ex. 2.14
Canonical variables, 55
Cantenoid, 95
Catenary (CVP 3), 32, 40,
  ex. 9.35
Cauchy-Schwarz
  inequality, ex. 9.2
Central field, 199
Chain rules, 3–4
Chaplygin's problem, ex. 3.30

Configuration variables, 55
Conjugate point, 171
Constrained problems, 67ff
Constraints
  holonomic, 78
  integral, 67
  nonholonomic, 81
  nonintegral, 78
Contour surface, 5
Contraction mapping princ., 230
Control
  bang-bang, 116
  feedback, 142
  optimal, 109ff
  signal, 112
Controlability, ex. 7.35
Convexity, 9ff,
Convex function, 9
Convex hull, ex. 1.36
Convex set, 9
Corner point, 209ff
Cost function, 17, 45
Cost functional, 161
  improper integral, ex. 8.33
Critical curve, 36, 47
Cruise-climb (CVP 5), 34
Curvature,
  constant, 73
  mean, 93
  normal, 94
CVP 1: 31, 47, 168
CVP 2: 32, 48
CVP 3: 32, 40, 49, 194
CVP 4: 33
CVP 5: 34
CVP 6: 35, ex. 4.22
CVP 7: 37, 169, 195

---

[1] Exercise 9.30

CVP 8: 37
CVP 9: 38
CVP 10: 39, 67, 71
CVP 11: 53, 57
CVP 12: 72
CVP 13: 75
CVP 14: 80, ex. 3.24
CVP 15: 91
CVP 16: 95
CVP 17: 96
CVP 18: 97
CVP 19: 100
CVP 20: 102
CVP 21: 102
CVP 22: 103

D

Damping,
    critical, 155
Degrees of freedom, 36
    multiple, 76
Dido's problem (CVP 10), 39, 67, 71
Dido, Queen of Carthage, 39
Differentiable, 2
Directional derivative, 1
Dirichlet
    inner product, ex. 6.14
    norm, 97, ex. 9.2
    principle, ex. 6.33
Divergence theorem, 92, 233
Docking, 116, ex. 7.21
Dual problem, ex. 2.15
Du Bois-Reymond, 59, ex. 4.35, ex. 11.18

E

Earth-Moon system, ex. 3.11
Eigenvalue, 98
    generalized, 101
    problems, 98ff
Electron, 97 (CVP 18)
Energy
    kinetic, 33, 52
    potential, 32, 52
    total, 54
Entropy, ex. 1.40
Erdmann corner conditions,
    first, 211ff
    second, 217ff
E-trick, 231
Euler-Lagrange equation,
    constrained problems, 69ff
    differential form, 45ff, 46
    integral form, 59
Excess function, 192ff, 213
Exponentially stable, ex. 8.30
Extremal
    curve, 59, 210
    regular, 217
    surface, 91ff
Extreme point, ex. 6.27,
    ex. 7.26, 119
Extremize, ex. 6.33

F

Feedback
    closed-loop, 142
    constant gain, 143
    control, 142
Fermat's principle, ex. 3.5
Feynman, ex. 1.40
Figurative, 213
Flows,
    stable (CVP 17), 96
    ODE, 186ff, 196
Force field, 77
    conservative, ex. 5.34
Functional, 162

G

Gradiant, 2, 21
Generalized
    eigenvalue problem, 101
    momenta, 54, 190

H

Hamiltonian, 54, 120, 190
Hamilton's equations, 55
Hamilton-Jacobi equation, 188, 190
Hamilton's principle, 36ff,
    with constraints, 81
Hermitian operator, 98
    see self-adjoint
Hilbert, 211
    inner product, ex. 6.14
    invariant integral, 191
Holonomic, 78

INDEX 253

I
Implicit function theorem, 226
Inclined plane, 82
Inner product, 11, 98, ex. 9.2
    Dirichlet, ex. 6.14
    Hilbert, ex. 6.14
Inverse function theorem, 68, 223
Isochrone
    Huygens, ex. 7.56
    Leibniz, ex. 7.57
Isoperimetric problems, 67ff
    catenary, 40
    Dido, 39

J
Jacobian, 8, 223
Jacobi
    equation, 171
    necessary condition, 212

K
Kinetic energy, 33, ex. 6.14

L
Lagrange multiplier, 7
Lagrange equations, 52
Lagrangian, 36
Laplacian, ex. 1.16
Least action, ex. 3.22
Legendre
    polynomials (CVP 20), 102, ex. 6.19
    condition, 174, ex. 9.26,
    strong condition, ex. 9.27
Lévy's theorem, ex. 2.30
Linear programming, 22
LQ problem, 139ff
LQR problem, 146ff

M
Mass-spring system, 37
    (CVP 7), ex. 3.7
Mathematical programming, 17
MATLAB cookbook, 235ff
Maximal principle
    Bauer, ex. 7.29
    heat equation, 93
Maximum principle
    Pontryagin, 120

Maxwell's equations, ex. 4.51
Mean value theorem, 5, 228
Minimum resistance (CVP 6), 35, ex. 4.22
Mirrlees, J. A., ex. 8.34
Momenta, 54, 190
Monte Carlo, exs. 2.17–2.32, 238ff
Moon landing, ex. 7.20

N
Natural boundary condition, ex. 4.36–4.37
Newton, I., see CVP 6
Normal system, 141
Normed vector space, 162ff
Norms, 162
    Dirichlet, 97, ex. 6.14, 169
    dominates, ex. 9.20
    Hilbert, ex. 6.14
No x result, 48, ex. 4.48
    generalization, ex. 4.53
No t result, 53
No y result, 48
No y' result, ex. 4.16

O
ODE, ordinary differential eqn.
OCP 1: 109, 125
OCP 2: 113, 124
OCP 3: 115
OCP 4: 116
Operations research, 17
Operator
    Hermitian, 98
    self-adjoint, exs. 3.15–3.18, 98, exs. 6.15–6.18
Optics, ex. 3.5
Optimal control, 109ff
Optimality principle, ex. 7.44
Optimization, 17ff
Orthogonal decomp., ex. 2.31

P
Partition function, ex. 1.40
PDE, partial differential eqn.
Pendulum
    articulated (CVP 9), 38
    elliptical, ex. 5.24
    extensible (CVP 11), 53, 57

planar (CVP 8), 37, 83,
  ex. D.38
 spherical, 81
Performance index, 113
Perturbation lemma, 231
Picard's theorem, 229
Piecewise
  continuous, 120
  continuously differentiable, 209
Plateau's problem, 94
POD, ex. 2.31
Pontryagin maximum princ., 120
Positive definite, ex. 3.15, 140
Potential energy, 32, 36
Principal solution, 171

Q
Quadratic form, exs. 9.7–9.15
  positive definite, ex. 3.16, 140

R
Rayleigh-Ritz numerics, 99ff
Reachable states, ex. 7.35
Reinvestments (OCP 2), 113
Riccati equation, 150, 170, 177,
Rolling cart (OCP 1), 109, 116

S
Schrödinger's eqn.(CVP 18), 97
Scripts
  Mathematica, 95, 187
  MATLAB, 233
Self-adjoint, exs. 3.15–3.18, 98, exs.
  6.15–6.18
Sensitivity, 3
Separation theorem, 11
Soap films, 91
  bubble, 95
Spring-mass sys.(CVP 7), 37
Stable
  asymptotically, 143
  exponentially, ex. 8.30
  flows, 96 (CVP 17)
  orbit, 97 (CVP 18)
State
  feedback, 142, 153
  space, ex. 7.12
Stationary value, 36
Stokes, 191

Streamlines, 187
Sublevel set, 5
Sufficiency
  strong, 185ff
  theorem, 174, 192, 200
  weak, 161ff
Switching times, 124, 126–128

T
Tangent plane, 6
Target, 112
Tautochrone, see Isochrone
Temperature, ex. 1.40
Transversality, exs. 4.36–4.37, exs.
  6.7, 7.31, 7.51

U
Utility, ex. 8.34

V
Variables,
  canonical, 55
  configuration, 55
Variational problems, 24
Variation
  first, 163
  second, 164
Vector space
  normed, 162
  topological, ex. 7.26
Voltage, 115

W
Wave equation, exs. 1.10–1.11
Weak minimum, 162
Weak*-topology, 119
Weierstrass
  E-function, 192, 213
  necessary condition, 214
Work, 77, ex. 5.30, 109
Wronskian, 171, ex. 9.43

Y
Young's double slit, ex. 2.32

Z
Zenodoros's problem (CVP 13), ex.
  3.27, 75